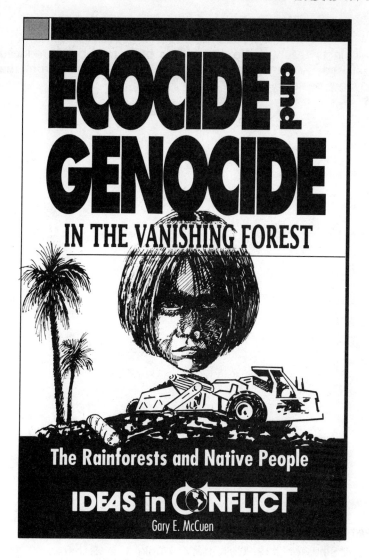

ECOCIDE and GENOCIDE

IN THE VANISHING FOREST

The Rainforests and Native People

IDEAS in CONFLICT

Gary E. McCuen

GEM
GARY McCUEN
publications inc.

411 Mallalieu Drive
Hudson, Wisconsin 54016
Phone (715) 386-7113

Illustrations & Photo Credits

Carol*Simpson 112, Cullum 47, Hubig 102, Ramirez 90, Ron Swanson 12, 19, 29, 41, U.N. Food and Agriculture Organization 35, 50, 56, 63, 74, 79, 84, 97, 108, 116, 122

© 1993 By Gary E. McCuen Publications, Inc.
411 Mallalieu Drive, Hudson, Wisconsin 54016

(715) 386-7113

International Standard Book Number
0-86596-087-9 Printed in the United States
of America

CONTENTS

CHAPTER 3 SAVING THE FOREST: IDEAS IN CONFLICT

REASONING SKILL DEVELOPMENT

These activities may be used as individualized study guides for students in libraries and resource centers or as discussion catalysts in small group and classroom discussions.

IDEAS in CONFLICT

This series features ideas in conflict on political, social, and moral issues. It presents counterpoints, debates, opinions, commentary, and analysis for use in libraries and classrooms. Each title in the series uses one or more of the following basic elements:

Introductions *that present an issue overview giving historic background and/or a description of the controversy.*

Counterpoints *and debates carefully chosen from publications, books, and position papers on the political right and left to help librarians and teachers respond to requests that treatment of public issues be fair and balanced.*

Symposiums *and forums that go beyond debates that can polarize and oversimplify. These present commentary from across the political spectrum that reflect how complex issues attract many shades of opinion.*

*A **global** emphasis with foreign perspectives and surveys on various moral questions and political issues that will help readers to place subject matter in a less culture-bound and ethnocentric frame of reference. In an ever-shrinking and interdependent world, understanding and cooperation are essential. Many issues are global in nature and can be effectively dealt with only by common efforts and international understanding.*

Reasoning skill *study guides and discussion activities provide ready-made tools for helping with critical reading and evaluation of content. The guides and activities deal with one or more of the following:*

RECOGNIZING AUTHOR'S POINT OF VIEW

INTERPRETING EDITORIAL CARTOONS

VALUES IN CONFLICT

WHAT IS EDITORIAL BIAS?

WHAT IS SEX BIAS?

WHAT IS POLITICAL BIAS?

WHAT IS ETHNOCENTRIC BIAS?

WHAT IS RACE BIAS?

WHAT IS RELIGIOUS BIAS?

*From across **the political spectrum** varied sources are presented for research projects and classroom discussions. Diverse opinions in the series come from magazines, newspapers, syndicated columnists, books, political speeches, foreign nations, and position papers by corporations and nonprofit institutions.*

About the Editor

Gary E. McCuen is an editor and publisher of anthologies for public libraries and curriculum materials for schools. Over the past years his publications have specialized in social, moral and political conflict. They include books, pamphlets, cassettes, tabloids, filmstrips and simulation games, many of them designed from his curriculums during 11 years of teaching junior and senior high school social studies. At present he is the editor and publisher of the *Ideas in Conflict* series and the *Editorial Forum* series.

CHAPTER 1

TROPICAL FORESTS:
AN ENDANGERED SPECIES

1   TROPICAL FORESTS: AN ENDANGERED SPECIES

THE VANISHING FOREST
Editor's Overview

Points to Consider:

1. How does the destruction of the rain forests affect the industrialized nations?

2. How much of the rain forests are already destroyed?

3. What is biodiversity? Give an example.

4. What is Japan doing to the rain forests of southeast Asia?

5. Can the rain forests be both exploited and preserved?

The world's tropical rain forests continue to disappear at a rate of nearly 2,500 acres every hour!

The year was 1987, the place. . .Brazil's Rondonia State in the Amazon rain forest. By September, nearly 8,000 fires were reported as trees were set ablaze by eager settlers homesteading the land. Futile attempts to transform the rain forest into farmland only ended in a scorched and barren land, displacing native people, destroying rare species of plants and animals and raising concerns over global oxygen production and climate change.

Global Ecocide

And the Amazon region is not alone—each year from South and Central America to Africa and Asia, nearly 2.5 million acres of rain forest are permanently destroyed. It amounts to ecocide on a massive global scale. Across the planet, the same economic and political forces are behind this eco-catastrophe. Millions of peasants trying to eke out a living slash and burn the forest, and soon discover that the soil can sustain their crops for only a few years. Greedy timber companies clear-cut the trees as a cheap source of raw materials. Laborers flock to gold mines in search of fortune, only to help destroy more forest and poison the land and water. Massive power projects dam the rivers and further threaten this global resource.

International banks and Third World leaders pursue extravagant development plans that lay waste to huge tracts of the tropical forest. The World Bank is currently preparing a $30 million loan for Cameroon in west-central Africa to be used primarily for financing a massive commercial logging operation that will affect 8.6 million acres of forest. This will spell doom for the 50,000 native Pygmies who have lived there for centuries.

Native People

Unable to easily acquire title to their own land, native people in every portion of the global rain forest are their now at risk. Their trees burned, their water poisoned and their food supply destroyed, these people are the victims of genocide in the vanishing forest. From the Yanomami of Brazil to the Mbuti of Africa and the Onge of India, entire cultures are threatened by disease, starvation and murder.

The Tukano of the Amazon continue to resist encroachment by the Brazilian army which has confiscated some 60 percent of their land for the Calha Norte Project — a scheme to secure Brazil's northern borders. The land has been designated "national forest" suitable for mining, logging and military maneuvers. Against all odds, the Tukano are fighting to block the construction of roads and barracks in their villages.

Meanwhile, the efforts of native Brazilian rubber tappers to cultivate the rain forest were set back by the tragic murder of their leader, Chico Mendes. Their vision of a sustainable, renewable rain forest have been challenged by wealthy land speculators in the Amazon.

Biodiversity

The tropical forests of planet Earth are essential to the stability of global climate and of our fresh water supply. Water stored in the thick vegetation of the tropics is evaporated back into the atmosphere, preventing droughts and floods. The trees also use carbon dioxide and produce oxygen, thus reducing what is known as global warming, and the greenhouse effect.

For millions of years, these great forests have evolved into the planet's largest variety of plant and animal life. It was only during the last 100 years that half of the world's original 7.7 million square miles of rain forest has been destroyed. While covering only seven percent of the earth's land surface, the rain forests are home to over half of all species, some 10 million in number. Many medicines and vaccines have been utilized for centuries by the native Indians, and countless more await

TROPICAL

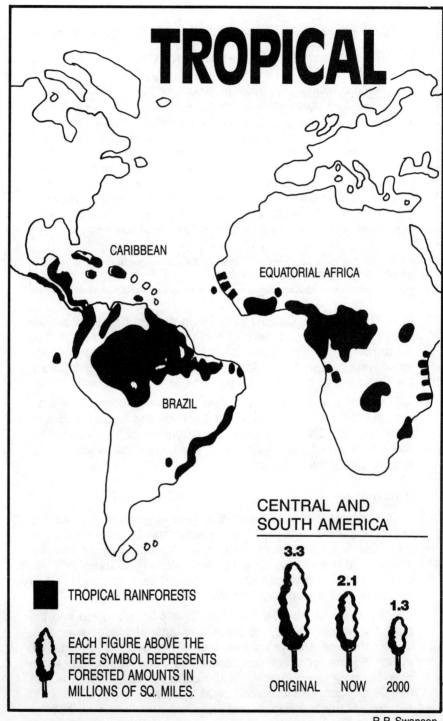

CARIBBEAN

EQUATORIAL AFRICA

BRAZIL

TROPICAL RAINFORESTS

EACH FIGURE ABOVE THE
TREE SYMBOL REPRESENTS
FORESTED AMOUNTS IN
MILLIONS OF SQ. MILES.

CENTRAL AND SOUTH AMERICA

3.3

2.1

1.3

ORIGINAL   NOW   2000

R.P. Swanson

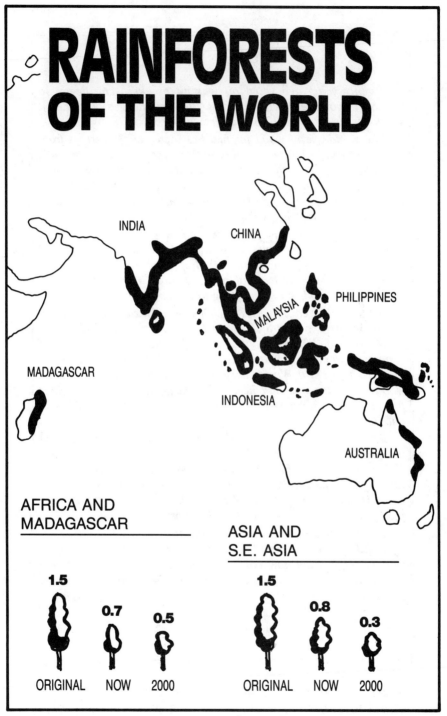

RAINFORESTS
OF THE WORLD

INDIA

CHINA

PHILIPPINES

MALAYSIA

MADAGASCAR

INDONESIA

AUSTRALIA

AFRICA AND MADAGASCAR

1.5
0.7
0.5
ORIGINAL   NOW   2000

ASIA AND S.E. ASIA

1.5
0.8
0.3
ORIGINAL   NOW   2000

Source: Rain Forest Action Network

WOOD AS FUEL

Twenty years ago when oil was cheap, kerosene began to replace firewood in the Third World. But that changed once the oil shocks began after the 1973 Arab-Israeli war.

The world use of wood for fuel has increased 35 percent from 1976-1986, according to the World Resources Institute. By 1987, 2.5 billion people—half the world's population—were cooking and heating with wood. Their daily energy use equaled the 21 million barrels of oil pumped daily by the Organization of Petroleum Exporting Countries, according to the World Bank.

With Iraq's invasion of Kuwait, what was a regression from a modern technology now is about to become a rout toward devastation of the environment: the sharp rise in oil prices threatens to send hundreds of millions more people scouring for any scrap of wood, straw and dung to burn.

In industrialized countries such as the United States, only 20 percent of trees go for fuel, the rest to industry, according to the World Wildlife Fund. But that still adds up to four percent of America's primary energy use, nearly as much as nuclear power. In most of the Third World, usage rates are reversed and mainly wood serves to cook and heat. Over 95 percent of energy comes from biomass—mainly wood—in Nepal, Rwanda, Sudan, Ethiopia and Tanzania.

The Food and Agriculture Organization estimates 1.5 billion of the people who rely on fuelwood are overcutting forests. Already 125 million can't afford to buy it. Fuelwood shortages exist in Sudan, Ethiopia, Somalia, India, Pakistan, Bangladesh, Central America and parts of South America. By the end of the decade, more than 2.8 billion people could be short of fuelwood.

For hundreds of millions living in lands stripped barren years ago, shifting back to wood will lower living standards for people whose lifespans are now only 45 or 50 years. For the rest of the world, it will accelerate trends already threatening the global environment.

Ben Barber, "Increased Use of Wood for Fuel Spells Trouble for Environment," **The Washington Post**, 1990

discovery. One fourth of all prescribed drugs in the U.S. are derived from tropical sources. Yet this great "biodiversity" of living organisms is at risk. Along with the forests, thousands of animal and plant species are lost to extinction every year. At current rates of destruction, we could be losing several species every hour! Some scientists fear that if deforestation continues unchecked, there may be no rain forest left by the middle of the next century.

The Rape of the Forest

Africa was the first continent to begin losing its rain forests when the colonial powers allowed private companies access to cheap timber supplies. Over 2000 square miles of rain forest were logged each year by European interests. This trend was continued by the post-colonial governments who were most eager to develop their newly independent nations with cash earned from wood exports. In addition to logging pressures, wood is also the primary source of fuel for heating and cooking in most developing nations.

The biggest threat to the rain forests of Southeast Asia is Japan's insatiable demand for wood products. Importing nearly 40 percent of the world's tropical hardwoods, Japanese companies are exploiting the forests of Indonesia and Malaysia with little regard for conservation. In China, over 300 animal species, including the giant panda, are threatened with extinction as a result of deforestation. And in Central America, logging, farming and years of civil war have destroyed 95 percent of El Salvador's original forest cover.

The Challenge

The population of tropical countries is expected to increase by 1.5 billion by the year 2000. As the amount of arable land becomes more scarce, the rain forests will continue to be viewed as a valuable source of agricultural production. In addition, many of these nations are burdened with tremendous foreign debt and will look to their natural resources as a means to provide cash and build their economies. Saving the rain forests while ensuring the economic livelihood of these developing nations is one of the key challenges facing the world as we approach the 21st century.

Conflicting social values are at the heart of the rain forest dilemma. Any solutions must address the issue of land use and title. The governments would pursue continued development while the indigenous people would prefer a cessation of all encroachment and clear title to the forest. Conservationists recommend preserving the forests through the creation of a series of national and international parks, while the

GLOBAL PHARMACY

For 500 years, since the people of South America encountered Europeans on their soil, the global pharmacopoeia has been enriched by a number of important plant-derived medicines discovered and utilized by indigenous people.

The skeletal muscle relaxant d-tubocurarine is derived from an Amazonian arrow poison better known as curare, Chonodendron tomentosum. *The anti-malarial drug quinine, obtained from the bark of the several species of Cinchona trees, was first called "Indian fever bark".*

One of the world's most important local anesthetics, cocaine, is derived from the leaves of Erthroxylum coca *and is still used today as medicine by thousands of people in the Andean region of South America. Pilocarpine, a drug used to treat glaucoma, is derived from the plant* Pilocarpus jaborandi *and was utilized by indigenous people in Brazil as medicine.*

Steven R. King, "The Source of Our Cures," **Cultural Survival Quarterly,** Summer 1991

rubber-tappers suggest a system of sustainable/renewable land use.

By alleviating Third World debt through an increase in aid programs, the industrialized nations could set a good example and encourage a more equitable system of land use that would benefit the local populations. Forest products such as silk, latex, materials for drugs and cosmetics, nuts, fruits, manioc, cacao, soybeans and coffee could feed millions and bring in cash without destroying the forests.

As the debate rages on, the world's tropical rain forests continue to disappear at a rate of nearly 2,500 acres every hour! If the ecocide and genocide is not halted soon, we may all lose one of our planet's most precious commodities forever.

2

TROPICAL FORESTS: AN ENDANGERED SPECIES

BRAZIL'S AMAZON: HISTORY AND FACTS

New Internationalist

The New Internationalist *is a monthly magazine published in Toronto, Canada, that deals with issues of world poverty and social justice. This reading was compiled from the May 1991 edition which dealt with the rain forests of Brazil's Amazon.*

Points to Consider:

1. When did the first humans arrive in the Amazon? When did the Europeans arrive?

2. What significance has rubber played in the history of the Amazon?

3. Why was 1987 such a crucial year for the rain forest?

4. What impact have dams had on the Amazon? Be specific.

Excerpted from, "The Amazon's Hidden History", and "The Amazon — The Facts", **The New Internationalist**, May 1991.

Brazil's Amazon rain forest contains 80,000 known, and at least 10,000 unknown, species of trees.

The Amazon's Hidden History

The people of the Amazon have left few historical records. This is because most of their artifacts were made from wood and other organic matter which would have rotted or disappeared. But attempts are now being made to reveal the history of the forest and its people.

Changing Forest: Recent scientific discoveries show that, far from being "ageless", the Amazon rain forest has undergone dramatic natural transformations. The most notable were during the Ice-Age. At this time the world's tropical regions became cooler and drier. The forest shrank and broke up and the savanna grasslands expanded. The small patches of forest or "refugia" that remained did not all evolve in the same way, with the same vegetation or animal life. So, when the forest eventually came together again there was great genetic variety within it.

Arrival of Humans: Humans are only recent arrivals in the forest. They first crossed the frozen Bering Straits into North America about 30,000 years ago—reaching the lowlands and forests of Latin America between 15,000 and 20,000 years ago. People first started farming and settling on the Amazonian floodplains some 5,000 years ago. Few settlements were static: groups would migrate long distances through the forest, little of which was left completely untouched. Conflicts between groups were regular occurrences. The more archaeologists discover about the forest, the clearer it becomes that estimates of how many people were living in the Amazon before the arrival of Europeans will have to be revised upward—to perhaps as many as 15 million.

European "Discovery": Soon after Columbus first set foot on American soil, Pope Alexander VI divided the uncharted lands of the "New World" between Spain on the Pacific coast and Portugal on the Atlantic coast. This happened at the Treaty of Tordesillas in 1494. Quite coincidentally, most of the Amazon Basin fell within the area designated for Portuguese colonization. Initial contacts between Portuguese explorers and the Indians were fairly friendly. They focused on the extraction of Brazilwood—used to produce dye—after which the Portuguese "colony" was named. The people were, of course, named "Indians" by Europeans because of their mistaken belief that they

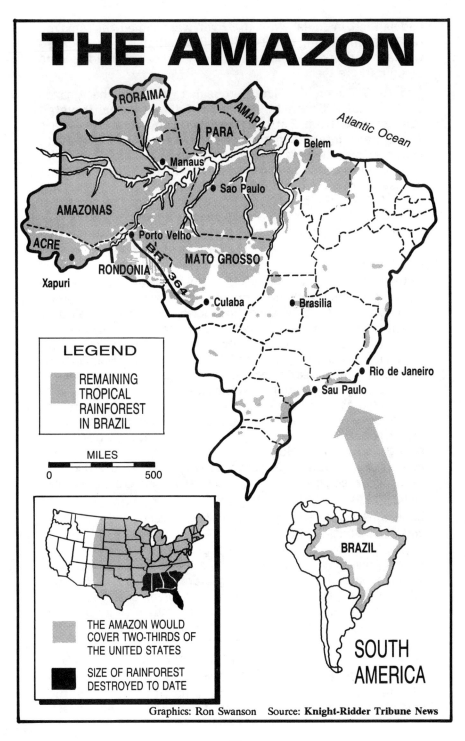

THE AMAZON

RORAIMA

AMAPA

PARA

Atlantic Ocean

● Belem

● Manaus

AMAZONAS

● Sao Paulo

ACRE

● Porto Velho

BR 364

MATO GROSSO

RONDONIA

Xapuri

● Culaba

● Brasilia

● Rio de Janeiro

● Sau Paulo

LEGEND

REMAINING TROPICAL RAINFOREST IN BRAZIL

MILES

0   500

THE AMAZON WOULD COVER TWO-THIRDS OF THE UNITED STATES

SIZE OF RAINFOREST DESTROYED TO DATE

BRAZIL

SOUTH AMERICA

Graphics: Ron Swanson   Source: **Knight-Ridder Tribune News** had landed in India. Why the river was named the Amazon remains uncertain, but it seems likely that it was because women warriors resembling those in Greek mythology were thought to live there.

Red Gold Rush: The first European to navigate the Amazon was actually Spanish. Francisco de Orellana traveled downstream from Peru in 1542. For the next 150 years Portuguese interest in the Amazon was largely limited to unsuccessful attempts to recruit Indian labor ("red gold") for sugar plantations on the coast. But indigenous peoples were not interested in working as wage laborers and violent conflicts ensued. The only Europeans to actually explore the rain forest were Christian missionaries—particularly the Jesuits. But in 1777 the first systematic attempt to develop the region was devised by the Portuguese Marquis of Pombal, mostly out of fear of encroachment by the Spanish, Dutch and British.

Rubber Barons: A small trade in rubber had already begun during the eighteenth century; by 1800 Belem was exporting 450,000 pairs of rubber shoes to England. But it was only after Charles Goodyear accidentally discovered "vulcanization" in 1842, and the industrial revolution increased demand for rubber products, that the Brazilian "boom" got underway. Rubber trees existed nowhere else in the world. Commercial houses, initially financed by the British, extended credit to laborers who penetrated the furthest reaches of the Amazon in search of rubber trees. Once there, the seringueiros (rubber tappers) were ensnared in debt bondage to the estate owners, the seringalistas, who sold basic necessities to them at grossly inflated prices. The trade collapsed during the First World War with the development of rubber plantations in Asia from plants smuggled out of the Amazon.

Land of Dreams: The great naturalists Alexander von Humboldt and Aime Bonpland traveled to the Amazon in 1799. This was to inspire botanists and explorers throughout the nineteenth century to roam the forest—to the displeasure of the Portuguese prior to Independence in 1821. These explorers returned with fantastic tales. The Amazon became the focus of nineteenth century romantic interest with the notion of the "natural state" and "the noble savage". Mark Twain wrote: "I was fired with a longing to ascend the Amazon. Also with a longing to open up a trade in coca with all the world. During months I dreamed that dream."

Empire of Schemes: The U.S. Navy conducted the first survey of the navigability of the Amazon in 1849. The first steamboats—which made it easier to ascend the river against the current—began operating in 1853. Growing U.S. interest in the Amazon found one expression in U.S. involvement in the revolution in the rubber-rich region of Acre, which in 1899 declared independence from Bolivia and finally became part of Brazil. The first of many U.S. businessmen to devise grandiose schemes in the Amazon, Percival Farquhar, managed to raise $70 million in Europe for a variety of projects, including the completion of the Madeira-Mamore railway in the middle of the jungle. It cost 6,000 lives to construct. Farquhar was ruined by the collapse of the rubber boom.

Military Maneuvers: Since the 1930s—and the "New State" established by the military President Getulio Vargas—the modern invasion of the Amazon has progressed along largely "strategic" and "geopolitical" lines. Particularly since the military coup in 1964, the "incorporation" of the Amazon into Brazilian territory has been the main motive behind government policies encouraging the colonization and deforestation of the area. Thus much of the initial deforestation in Rondonia and Para took place around "Development Poles" constructed by the government from 1966 onward. These were combined with a preference for "big projects", building roads, dams and other debt-inducing industrial enterprises largely with the backing of multinational lending agencies like the World Bank.

The Amazon — The Facts

The Amazon rain forest is a living thing. It lives in balance—or "equilibrium"—consuming what it produces. If the balance is lost it may die.

The Facts of Life. . .

THE RIVER

- is the largest river system in the world: four times larger than the Zaire (the second largest), eleven times larger than the Mississippi.
- disgorges 198,000 cubic meters of water per second—enough to fill Lake Ontario in three hours.
- contains one fifth of the world's fresh water—or two-thirds excluding water locked in polar ice caps.
- flows a distance of 6,762 kilometers from its source in the Peruvian Andes to its mouth—equal to the distance between London and New Delhi.
- has 10,000 tributaries totalling 80,000 kilometers in length

which would stretch twice around the Equator.

- provides 24,000 kilometers of navigable "trunk" waterway — ocean-going ships can penetrate up the Amazon a distance equivalent to crossing the North Atlantic.

THE FOREST

- extends over five million square kilometers — that is 10 times the size of France.
- makes up one-third of the world's remaining tropical rain forest.
- has 30 percent of all known plant and animal species.
- contains 80,000 known, and at least 10,000 unknown species of tree.
- has a density of between 100 and 300 tree species per hectare of forest (temperate forests have between five and ten.)

THE ANIMALS

In the Amazon there are:

- one fifth of the world's bird species in scarcely one fiftieth of its land surface.
- several million animal species, mostly insects: one tree stump in Bolivia was found to house more ant species than the whole of the United Kingdom.
- 3,000 known species of land vertebrates.
- 2,000 known species of fresh water fish, or ten times as many as in the whole of Europe.

THE ENVIRONMENT

The Amazon Basin (the area drained by the river)

- covers some 7.5 million square kilometers — an area almost as big as Australia — in six different countries (Brazil, Bolivia, Peru, Ecuador, Colombia, Venezuela).
- is the wettest region on earth, with an average rainfall of 2.54 meters per year.
- is surrounded by one of the youngest rock formations on earth (the Andes Mountains) to the west and two of the oldest (the Guyana and Brazilian Shields) to the north and

south.

- has very poor soil; 90 percent suffers from phosphorous deficiency, 50 percent from low potassium reserves and 24 percent from low drainage or flood hazards.

The Facts of Death

DEFORESTATION

- As much as 75 percent of deforestation in the Amazon has resulted directly or indirectly from large-scale agricultural or industrial schemes. Many of them have received funding from international agencies like the World Bank.
- Estimates of the total amount of forest now cleared vary between five percent and 20 percent. Most independent experts now accept a figure of 12 percent by 1985, of which 75 percent has taken place since 1960.
- In 1987, probably the worst year for deforestation so far, satellites detected 8,000 fires in the states of Rondonia and Mato Grosso between June and September. In that year, an area of 210,000 square kilometers, almost as big as Great Britain and Ireland, was cleared.

FLOODING BY DAMS

- Brazil's "2010 Plan" envisages the construction of 31 hydroelectric dams in the Amazon Basin.
- Two dams have already been completed: Tucurui in Para and Balbina near Manaus have flooded a total of some 5,000 square kilometers of rain forest.
- Much of the energy produced by Tucurui is consumed by aluminum smelters. The construction of the Balbina Dam cost $700 million, and a further $700 million may be needed to increase its efficiency.
- The Waimiri-Atroari Indians living in the neighborhood of the Balbina Dam have been decimated both by the dam and by the BR 174 Highway running north: in 1972 they numbered 3,000, but by the mid-1980s their numbers had been reduced to less than 300.

HIGHWAYS TO HELL

- The Calha Norte ("Northern Trench") project was begun by the military in 1985 to develop the region's network of roads, increase the military presence in the area and improve the demarcation of national boundaries.

- It covers 24 percent of Brazilian Amazonia in a 6,000-kilometer long corridor which includes 84 indigenous areas, 51 different Indian people, and 557 mining claims.

- Official policy is to promote the colonization of Indian people, converting them to the life of agricultural peasants on one-square-kilometer size plots.

UNDERSTANDING MAPS

Maps can provide us with a graphic understanding of the political and social issues facing our world today. The purpose of this activity is to familiarize the reader with the geographic extent of the global rain forests, and their demise.

Guidelines

1. Referring to the map on pages 12 and 13, what area has the most remaining rain forests?

2. How many square miles of original rain forest once existed (total) between the three areas listed? How much is left in these regions today?

3. Referring to the map on page 19, how much of the Amazon rain forest has been destroyed? Give your answer in terms of U.S. states.

4. Using the scale of miles on the maps, determine the approximate distance across the Amazon rain forest from the towns of Xapuri to Belem.

5. Locate an atlas or globe and find your own city or town. Using the distance you have just calculated, determine how far you could travel from your home in any direction across land. This will give you an idea of the size of Brazil's Amazon rain forest.

CHAPTER 2

RAIN FORESTS UNDER SIEGE

3   RAIN FORESTS UNDER SIEGE

LABOR RIGHTS AND ECOLOGY IN THE AMAZON

Samantha Sparks

Samantha Sparks is a contributing writer for the Multinational Monitor. *She wrote this article on labor rights for the rubber tappers of the Amazon.*

Points to Consider:

1. Who was Francisco "Chico" Mendes?

2. Why are rural workers in the Amazon being killed? Who is responsible?

3. Which group has a legal right to the land? Why?

Samantha Sparks, "Labor Rights and Ecology in the Rain Forest", **Multinational Monitor,** March 1989.

While local landowners want to clear the forest for cattle grazing, logging and land speculation, the rubber tappers union is fighting to keep it intact.

When Francisco Mendes Filho was assassinated in the Brazilian Amazon in 1988, U.S. environmentalists responded with a barrage of protests. Mendes, a leader of a local rubber tappers' union, had become a powerful symbol in the battle to halt destruction of the world's largest rain forest. What many of the eulogies for Mendes missed, however, was that his was first and foremost a trade union struggle—a fight to save jobs and protect workers' rights in the Amazon. Saving trees came second. "We became ecologists without even knowing the meaning of the word," Mendes told the *New York Times.*

Labor and Environment

"Chico Mendes didn't start as an environmentalist," said Stan Gacek, assistant director of international affairs for the United Food and Commercial Workers (UFCW). "He started as a trade unionist. But Chico Mendes, probably more than any other Brazilian unionist, linked the issues of labor and the environment."

The UFCW and Mendes' union are loosely joined through their affiliation to the International Federation of Plantation and Agricultural Workers. And the UFCW has maintained a "fraternal relationship" with Mendes' Rubber Tapper Union of Xapuri for several years, says Gacek, It was Mendes' union work that prompted the AFL-CIO and several U.S. unions to send telegrams demanding investigation of the Mendes killing.

Mendes had also garnered support from many quarters outside of organized labor. Senator Bob Kasten (R-WI), known for his anti-union track record in the U.S., publicly mourned the death of the union leader. And the Environmental Defense Fund, a U.S. environmental organization, established a fund to continue the unionist's work. The EDF, however, did not appoint a single U.S. labor representative to the fund's board.

Although Mendes' death has attracted unprecedented attention, it was not the first of its kind. In 1988 alone, 95 rubber tappers were murdered, according to the church-based Pastoral Land Commission, says Gacek. During a trip to Washington, Raimundo de Barros, who replaced Mendes as president of the Xapuri union, said that he is, as Mendes was, on a death list. Since 1980, more than 1,500 assassinations have taken place in

rural areas; about one half of these occurred after 1985, when a civilian government replaced a 20-year military dictatorship. "In the rural sector, human rights have gotten worse" since the new government, pledging democracy, took power, says Gacek.

Demanding Their Rights

The reason for the killings is simple: rural workers have begun demanding their rights to land that cattle ranchers and other investors want for themselves. Mendes was shot, local sources charge, on the orders of Darly Alves, a cattle rancher in Acre, Mendes' home state. The cattle ranchers feared that Mendes and his union were beginning to gain control of forest land. His death was not unexpected: he had been under police protection for more than a year because of repeated death threats. Less than three weeks before his death, Mendes denounced the Acre State Police and cattle ranchers in a national newspaper for conspiring to kill him.

HIGH STAKES

As he walked to the bathhouse in his back yard in Xapuri, in Brazil's western Amazon, Brazilian rubber-tapper Francisco Mendes Filho, 44, was assassinated by a point-blank shotgun blast to his heart. The world-renowned ecologist and inspired leader of the rubber-tappers' movement was murdered.

It was no small effort to organize the rubber-tappers, scattered in their remote settlements. The union in Xapuri became the nucleus of resistance to the invasion of lands occupied by tappers, Indians and peasant farmers in Acre, Brazil's westernmost state in the Amazon basin. Coping with death threats and assassinations was a tragic aspect of the union's function.

The treacherous death of Chico Mendes dramatizes the social conflict that underlies Amazon development and deforestation.

Marianne Schmink, "Murder Illustrates Stakes in Struggle for Rain Forests", Newsday, March 12, 1989

While local landowners want to clear the forest for cattle grazing, logging and land speculation, the rubber tappers union is fighting to keep it intact. Members' jobs depend upon the rain forest's survival. "It's very much a class issue," says Gacek. "You have a labor/capital struggle with the rubber tappers trying to preserve their livelihood. The ruling classes feel more threatened and are reacting more violently."

Mendes was one of about 500,000 rubber tappers remaining in Brazil. Most of them are descendants of migrants who came to the forest during the rubber boom of the late 1800s. The rubber tappers have never had it easy: from the very first they were cheated by avaricious landowners and middlemen who taxed them heavily for their use of the forest and bought their rubber at much lower prices than its market value. Today, rubber tappers may appear to be a dwindling group plying an archaic trade, but in fact, domestic rubber continues to play an important role in the Brazilian economy.

Government Policies

For over a century, Brazilian governments have recognized this role, but until recently government policies favored the landowners who profited from the rubber tappers' work. A state monopoly guaranteed landowners a minimum rubber price, and imported rubber was taxed to make it cost as much. Over the past decade, however, national rubber policy has begun to strengthen the rubber tappers' hand. According to EDF anthropologist Steven Schwartzman, in 1980 the government began to encourage the rubber tappers to sell their product directly to industry, for higher prices than they could get from the middlemen.

Other recent developments have tipped the traditional balance of forces in the Amazon more abruptly. Most importantly, ownership of the forest has changed dramatically. As a result of a state government drive to attract investment and develop the region, land values shot up more than 1,000 percent during the 1970s. Most of the old landowners sold their estates to wealthy investors from outside the Amazon states. The newcomers quickly realized that the rubber tappers and the forest stood between them and the faster and bigger profits they could earn by clearing the land and opening it up to cattle ranching.

What the new investors ignored, however, was a Brazilian law enabling anyone who occupies land for more than a year and makes improvements on it to lay claim to that land. Thus, the rubber tappers, as squatters of sorts, could claim legal rights to the land. When they demanded these rights, the bloody battles ensued. In a recent paper Schwartzman writes that, "tactics used to evict rubber tappers and clear land for sale included threats, violence, burning of houses and destruction of crops." The union movement which Mendes led since 1980 was born of the rubber tappers' resistance to these attacks.

Unions

By 1980, rubber tappers had organized unions in more than 60 percent of the municipalities in the state of Acre, and a state federation of unions was formed. There are now an estimated 30,000 union members in the state, of whom about 80 percent are rubber tappers.

In part because of a strategic alliance forged with U.S. environmentalists, rubber tapper unions in the Amazon have made impressive gains. Out of the spontaneous resistance to landowners' attempts to evict them some 20 years ago, the unions have developed a coherent strategy with a specific goal. They want substantial parts of the rain forest set aside as "extractive reserves" – places where the rural workers can continue to live and work by harvesting rubber and Brazil nuts without destroying the forest.

The rubber tappers have received support from U.S.

31

environmentalists and, increasingly, development institutions like the World Bank, because their call for extractive reserves makes long-term economic sense. Their plan preserves the resources of the forest instead of exploiting them for short-term profits.

At the urging of the rubber tappers' unions, 12 extractive reserves, totalling more than five million acres, in five Amazon states are being created, according to EDF. "The immediate goal of the rubber tappers is to set up reserves," notes Gacek. "But like all unionists, they have wider concerns, such as their rights as salaried workers, and rights for those who are small-scale landowners."

Despite their gains, the future of the rubber tappers' work appears uncertain at best, and it is likely that more people will die in the fight to protect workers' rights in the Amazon.

RAIN FORESTS UNDER SIEGE

GREED BREEDS ECOCIDE IN THE PHILIPPINES

Dan Connell

Dan Connell wrote this article about the threat posed to the Philippine rain forests by the military, the government and big business.

Points to Consider:

1. What is *kamagong*? Why is it significant?

2. Who is destroying the rain forests of Palawan?

3. How are the indigenous people affected?

4. Why is it dangerous to protect the forests?

5. What role does government corruption play?

Dan Connell, "Greed Breeds Ecocide in Philippine Paradise", **Guardian,** September 18, 1991.

Crushing poverty throughout the Philippines appears to be sparking the rush to exploit Palawan's remaining resources.

Oscar Lapida was "answering a call of nature" when earlier this year he wandered into the dense rain forest in southern Palawan and stumbled onto a secret cache of contraband *kamagong* logs worth nearly $1 million. As a community organizer for the Haribon Foundation, one of the Philippines' oldest and most respected environmental agencies, he sought, with difficulty, to notify government authorities.

"I was shocked—this was my first time to see so much *kamagong* in one place," said the 28-year-old native of Palawan, the Philippines' last true "frontier" and site of an escalating battle between conservationists and developers. "I tried to send a message from the local military detachment, but they refused to let me," Lapida recounted. "Later, I found out that they were protecting the logs."

Corruption

His report sparked a firestorm of charges and countercharges of corruption, cover-up and sabotage that has rocked this traditionally peaceful island, one of the few as yet untouched by the bitter 22-year civil war raging throughout the rest of the country. The issues raised by the incident also go to the heart of an intense national debate over the government's export-driven growth strategy. Critics charge that the country's once-abundant endowment of flora and fauna is being squandered by Manila while tens of millions of people are consigned to lives of unremitting poverty.

Conservationists accuse the local military command of illegally cutting the nearly extinct hardwoods—an extremely rare form of ebony known here as "black gold"—and trying to smuggle them to neighboring Malaysia. They also claim that powerful politicians are profiting from the destruction of Palawan's rapidly dwindling natural resources. Military and political leaders counter that the environmentalists are "communists", and have filed formal charges of subversion against 14 community organizers from Haribon and several local human rights agencies.

"It's not true," insists Flordelis "Bogs" Fernandez, the only woman among the 14 accused. "This is an effort to smear and discredit us." Military units have also allegedly attacked members of the Palawano tribe who live in the rain forest where the logs

Workmen clear virgin section of tropical forest on Mindinao in the Philippines. Source: U.N. Food and Agriculture Organization.

were discovered, claiming they are helping Haribon and also hiding guerrillas of the Communist-led new People's Army. A detachment of 30 marines beat several Palawanos, hung one by the neck for hours with his toes barely touching the ground and ransacked 50 houses during a recent raid, according to a report by Task Force Detainees, a national human rights monitoring group.

From Paradise to Wasteland

A two-week tour of Palawan revealed both the incomparable natural beauty of this remote Western Pacific island and the extent to which its primeval forests and stunning offshore coral reefs have already been destroyed by uncontrolled growth. The visit also showed the degree to which Palawan is polarized over environmental issues.

Palawan contains the last significant stand of rain forest in the Philippines and the richest fisheries in the sprawling archipelago. However, a flood of migrants from all over the impoverished island nation, coupled with the twin scourges of illicit logging and illegal "blast" and cyanide fishing, threaten to turn this tropical paradise into an arid wasteland by the turn of the century, according to Haribon.

There are still mountain valleys packed with impenetrable curtains of lush green foliage. Tangles of thick, gnarled gray vines spill out of majestic stands of "ipil," "almaciga" and "ironwood" trees, whose thin root "shoulders" start at heights of more than six feet and wind out from the trunk like twisted flying buttresses. Delicately patterned butterflies flit in and out of the pristine forest that is home to over 3,000 varieties of plants and scores of rare, endangered species of animals and birds, including the Palawan bear cat, the scaly anteater and the Philippine eagle.

Driving north through the island's largest logging concession, however, a visitor sees a patchwork of haphazardly cultivated valleys, scrub-covered hilltops and barren, rocky slopes that have been slashed and burned. The destruction permits "shift" farming by migrants, who move from one freshly logged site to another every three to four years.

Great clouds of bone-dry, ocher dust billow up behind the four-wheel drive vehicle that winds through what was once steamy jungle. Now only spindly, second-growth trees and tufts of brittle, brown grass grow here. Local farmers say that temperatures have skyrocketed without the forest cover, threatening a drought here for the first time in Palawan's history.

The Philippines Department of the Environment and Natural Resources estimates that loggers are responsible for 25 percent of the downed trees while the rest are destroyed by shift farmers. Haribon organizers argue, however, that the loggers open the forests to the migrant farmers by building new roads through the jungle. "If the logging stopped, we are sure that 75 percent of the destruction would stop also," said Oscar Lapida.

Crushing Poverty

Palawan is often called the "Little Philippines" because it is home to more than 20 ethnic groups. Three of these are indigenous, and they are now threatened with extinction by the disappearance of the forests, according to Haribon officials who said the tribespeople are an essential component of the rain forest ecosystem. "The tribes are the best protectors of the forest because they know it best and they need it for their own survival—without the forest, they are nothing," explained Lapida.

Palawan's other inhabitants come from all across the Philippine archipelago, which stretches north some 1,500 miles

from Malaysia almost to Taiwan. Palawan's jagged mountain
ridges, narrow coastal plain and extremely limited watershed
made it one of the last islands to attract large numbers of
settlers, though the earliest human remains in the Philippines,
estimated to be 40,000 years old, were discovered in southern
Palawan.

With five percent of the country's land mass, Palawan houses
only 750,000 people out of a total Philippine population in
excess of 60 million. Yet even this is triple the local population
of less than 30 years ago, and the numbers continue to grow at
an explosive annual rate of 4.6 percent, mainly from migration.

Crushing poverty throughout the rest of the country appears
to be sparking the rush to exploit Palawan's remaining
resources. By government estimates, over 60 percent of Filipinos
now live below the poverty line, and the Philippines is reported
to have the highest rate of malnutrition in Asia, including
Bangladesh.

Concessions and Corruption

The rampant destruction of the country's forests has provoked
a national debate over how and to what extent commercial
logging should be regulated. A groundswell is building for a
25-year logging ban while longer-term policy is hammered out.
Hasty efforts by Pagdanan Timber Products, the largest logging
company in Palawan, to grade the island's main logging road
and to build a string of wooden bridges and new feeder roads
prior to this past summer's rainy season suggest that the
company is trying to speed up its operations before the

politicians make up their minds. Meanwhile, government rents on the logging concessions, at only $60 per acre, are among the lowest in the world.

Pagdanan Timber is controlled by logging magnate Jose "Pepito" Alvarez, who got his start in the industry in Indonesia and is now reportedly opening operations in Vietnam. Alvarez has frequently been charged with running illegal logging operations, although he consistently denies doing so.

Threats and Murder

Under these conditions, efforts to protect Palawan's environment can be dangerous as well as controversial. A foreign priest working on rain forest protection issues was shot and killed. The leading suspect was a major commercial logger, but no formal charges were brought. After an expose in the *Far Eastern Economic Review* alleged a direct link between the murder and Alvarez, one of the authors was hit with a libel suit and received death threats. Even government officials are not immune to pressure, according to one who says she has been threatened with death and offered bribes of up to $1,000, matching her annual salary, to ignore environmental laws.

Meanwhile, 400 of the 500 *kamagong* logs originally discovered—and photographed—by Haribon organizer Oscar Lapida less than 100 yards from a Philippines marine camp have mysteriously disappeared. Haribon officials say they are probably in Malaysia by now. Military officials say they never existed.

5   RAIN FORESTS UNDER SIEGE

HAWAII – SPECIES EXTINCTION CAPITAL OF THE U.S.

Paul McEnroe

Paul McEnroe is a national correspondent for the Minneapolis-based Star Tribune.

Points to Consider:

1. Why is Hawaii called the "species extinction capital" of the U.S.?

2. What damage is done by the mongoose and the feral pig?

3. How will geothermal development affect the rain forest?

4. Why are native Hawaiians angry?

Paul McEnroe, "Hawaii's Rain Forest Faces Battle to Survive", May 9, 1990. Reprinted with permission of the **Star Tribune,** Minneapolis, St. Paul, Minnesota.

Seventy percent of the extinctions of plants and animals in the U.S. has taken place in Hawaii.

Tim Tunison likes to think of it as his ride through environmental hypocrisy. To his left along the narrow road bounding the park border is a great green wall, a rain forest he tries to heal with everything from hog dogs to herbicides, so thick it takes an entire day to struggle through two miles, a world nobody should wander into without reading the sun and backing it up with a compass, lest he wanders in circles for hours. On his right is his version of disaster in the name of progress. "Stop the car," he ordered, peering through his glasses fogged by a morning shower. He's staring at something he's never gotten used to as a botanist at Volcanoes National Park on the island of Hawaii.

Our Brazil

"This is our Brazil. This is our Costa Rica and Nicaragua," he says, pointing at the destruction. What was once a rain forest on his right is nothing but a vast junkyard, a pasture studded with a few trees and cows walking around rusting cars.

"In the '50s, the state auctioned off these 30-acre chunks of rain forest to be turned into agricultural lots and thus encourage development," Tunison says. "There's still a tax benefit if you turn rain forest on your property into grazing land. We got the part of rain forest land they didn't sell off and turned it into a park."

"The hypocrisy here is how can we as a country ask Brazil to stop the devastation of the rain forest in the Amazon basin if we can't even control what's left on a small island? Just five years ago this state was allowing rain forests on this island to be cut for wood chips as an alternative fuel source."

Now there's a new threat. Looming on the horizon just a few miles from where Tunison wages his small fight for preservation is the beginning of what may turn out to be unprecedented scarring of the Hawaiian landscape, a slash through the last significant rain forest in the United States, that is forcing Hawaiians to rethink their environmental values, religious beliefs and cultural identity as never before. It's called geothermal energy, and Hawaii's $8 billion-a-year tourist industry hungers for it.

A Broken Chain

The environmental chain in Hawaii, once the most isolated archipelago in the world with delicate links that were joined through millions of years of evolution, is broken beyond repair. Once, cactus grew there without spikes, flies without wings, mint plants lost their mint flavor and berry plants grew without thorns. There was no need for the evolution of strong defenses because plants and insects were not being attacked by any mammals or reptiles. Because of that isolation, more than 90 percent of Hawaii's indigenous plants and animals are found nowhere else on Earth. But, said Tunison, "Hawaii has become the species extinction capital of the United States."

The chain began breaking with colonial exploration—when

Captain Cook's crew in the late 1770s dumped ravaging pigs overboard to be used as food sources upon the sailors' return, and later in the 1880s when plantation barons imported Jamaican mongooses to keep the rats that had escaped from visiting ships from ravaging fields of sugar cane.

Today, the threat stems from six million tourists who visit annually, spawning demands for more hotel complexes and longer airport runways that would allow bigger planes to bring even more tourists.

"I'd call the situation a crisis, an aesthetic and scientific crime because unique things are being destroyed," said Faith Campbell, a rain forest specialist with the Natural Resources Defense Council. "The habitat destruction process has been allowed to go too far and the broader picture is we'll lose more." Less than a quarter of Hawaii's native forests still stand and only now has the state pledged to find permanent funding to protect what's left.

Mass Extinctions

"But when you only are left with small pockets of forest, it will be very tough to keep what's left intact," said Tunison, standing in a portion of park rain forest ravaged by wild pigs — descendants from the Cook expedition. "On an island ecosystem we are dealing with alien species management — over 50 alien species of plants and animals threaten what's left in the national park."

For the 30,000 acres of rain forest within Volcanoes National Park, there is only enough money to work on preserving 25 percent, said Tunison. Further, if there is geothermal drilling, the energy sites would ring a major portion of the park's border and allow further introduction of alien plants and animals.

Seventy percent of the extinctions of plants and animals in the United States has taken place in Hawaii. Yet because Hawaii is so far from the political decisions in Washington, only about two percent of the nation's endangered-species funding goes to Hawaii, Campbell said.

At least 50 percent of the native birds are extinct. Thirty of the 42 species that remain are listed as endangered. Much of that problem began with the introduction of the mongoose. Trouble was, the rats carried out their destruction at night, while the mongooses slept. Instead of rats, the mongooses went after ground-nesting birds, eating eggs and young adults. Today, the mongoose population is out of control.

Fifty percent of Hawaii's native plant life is on the federal government's endangered-species list. "People usually yawn when you talk about plants," Tunison said. "But what most don't realize is that without these plants you wouldn't have animals. You wouldn't have rain forests that make this place so unique."

"Do you notice? You can't hear any birds singing in here, can you? They've been wiped out. And we're standing in mud. The mosquitoes carrying avian malaria lay larva in the pools of water formed after the pigs virtually rototill an area to root out the tree fern. The birds, with no natural defense in their immune system—since they evolved with no need for it—are killed by malaria.

"All this light in here? The fern canopy turns rain into mist. Without the canopy, alien plants come in because there's more light—alien plant species like the banana poke vine climb the trees and kill them. When you think of it this way, you can understand why plants and birds are so important."

There was one distant sound, though. Down by the park's kennel, Koa, Gus, Freddie and the rest of the hog dogs were baying, eager to break out and pick up the scent of the feral pigs that continue to destroy the park's rain forest. "You've got to wipe out about 80 percent of the population a year for a couple years in order to do any good," Tunison said. "We set

43

snares, or the dogs track 'em and then we kill 'em."

Geothermal Threat

The idea is to put about one hundred 20-acre geothermal drill sites in the middle of the Wao Kele O Puna rain forest, the biggest remaining lowland forest of its kind in the United States, located just outside the national park boundaries. The project that would tap the energy of the still active Mauna Loa volcano would cost up to $1.5 billion and be borne by consumers. Sulfurous steam would be converted into electrical energy and be transferred by underwater cable to other islands, most notably Oahu, where 80 percent of the population lives.

Opponents say it literally will stink up the atmosphere, especially if there's an earthquake strong enough to knock drill caps off the wells. And it will mean the introduction of new animals and plant species that will attack the rain forest. The project is being held up by a case pending in the U.S. Court of Appeals on whether native Hawaiian rights and access to the forest are being violated.

Previously, the Supreme Court declined to hear a First Amendment claim brought by the Pele Defense Fund that freedom of religion was being denied because the drilling violated the native beliefs that the volcano goddess, Pele, was being desecrated.

"We are saying you can't ruin a rain forest and stomp on a culture without it being noticed by the rest of the country," said Nelso Ho, a member of the Sierra Club's board of directors who is a fourth-generation Hawaiian of Chinese descent.

"Geothermal will make this island the Pittsburgh of Hawaii because the sulfur gases they tap into will stink—not to mention causing more respiration problems for the children and elderly who already suffer from the sulfur in the air. The rationale is greater good for the greatest number—they want to transfer the electricity by underwater cable to Oahu while we're left on the Big Island with the smell and the continued destruction of a forest."

To Pali Kapu Dedman, a native Hawaiian and president of the Pele Defense Fund, the geothermal project represents the "plantation mentality" that he says permeates the state's political scene. "Almost everyone who immigrated here was brought in to work and there was no concern about looking out for the native people and their environment," he said. "As native people, we are not seen as part of the environment. Yet we are endangered—we have the lowest life expectancy on the islands."

44

"Nobody does a psychological impact study on what this will do to our culture and religion. You are fooling around with 1,000-year-old decisions about spirituality and deities and what Pele means to us. More scars across the rain forest. All those cuts will end up killing it. In the name of progress."

INTERPRETING EDITORIAL CARTOONS

This activity may be used as an individualized study guide for students in libraries and resource centers or as a discussion catalyst in small group and classroom discussions.

Although cartoons are usually humorous, the main intent of most political cartoonists is not to entertain. Cartoons express serious social comment about important issues. Using graphic and visual arts, the cartoonist expresses opinions and attitudes. By employing an entertaining and often light-hearted visual format, cartoonists may have as much or more impact on national and world issues as editorial and syndicated columnists.

Points to Consider:

1. Examine the cartoon on the following page.

2. How would you describe the message of the cartoon? Try to summarize the message in one to three sentences.

3. Does the cartoon's message support the author's point of view in any of the readings in Chapter Two of this publication? If the answer is yes, be specific about which reading or readings and why.

Cartoon by Cullum, **Copley News Service**

6   GENOCIDE AND NATIVE PEOPLE

INDIGENOUS LEADERS STRUGGLE FOR RESPECT

Marc Mowrey

Marc Mowrey is a free-lance writer based in San Francisco. He is currently working on a history of the modern environmental movement.

Points to Consider:

1. What is COICA?

2. What is a "debt-for-nature swap"?

3. Why is Peru exploiting its tropical rain forests?

4. What is happening to the Chimanes of Bolivia?

Marc Mowrey, "An Alliance of Equals", **Sojourners**, Aug/Sept. 1990. Reprinted with permission from **Sojourners**, P.O. Box 2972, Washington, D. C. 20017.

"We either disappear with the forest, or live with the forest. We have no other place to go."

Indigenous people have lived in the tropical rain forests of South America's Amazon basin region for millennia. And while their way of life has been threatened by outsiders (including Christian missionaries) for centuries, it is only in the last few decades that they have faced the threat of extinction due to development—i.e. gold prospectors, dams, cattle ranchers, oil wells, and lumber companies.

Saving the Forest

Saving the forest has become a popular cause in recent years, with presidents, kings, religious leaders, corporations, scientists, and environmentalists discussing Its fate. But in all of the discussions on topics such as global warming and the genetic diversity of the rain forest, one group has been conspicuously absent: the 1.5 million people who live there.

As part of an effort to place themselves at the bargaining table for their survival, indigenous leaders from Peru, Brazil, Bolivia, Colombia, and Ecuador—who are part of a coordinating body of indigenous organizations known as COICA—hosted environmentalists in 1990 for what was called the First Iquitos Summit, held in Iquitos, Peru.

With the help of environmental groups, the indigenous leaders hope to gain more control over the fate of their native homelands. And based on the historic document signed at the meeting's end, they are succeeding. The document formally declared a commitment among the indigenous people and the environmentalists to continue working together—as an alliance of equals.

A Matter of Land Titles

Specifically, indigenous people of the Amazon basin are seeking land titles. Most of the current territory they occupy is untitled, and as such is subject to the whims of the respective national governments in the region. If, for example, a timber company offers the minister of agriculture a sizable chunk of cash for logging rights to untitled Indian land, no legal barriers exist to prevent the selling of those rights.

Of course, government-issued titles are no guarantee in a country such as Peru where, according to Raymond Offenheiser of the Ford Foundation, "the central bank is flat on

49

Washing clothes in the Amazon Basin. Source: U.N. Food and Agriculture Organization

its back and the government cannot meet staff payroll." After all, land titles, like currency, are just pieces of paper—only as valuable as the bank behind them. However, they are a step in the right direction.

Land titles, while a priority, are just one of many demands the indigenous people are making in an attempt to secure their homelands for future generations. They also want research done on indigenous human rights violations, an end to Indian slavery in Brazil and Peru and to forced displacement, official recognition of tribal governments, and no multilateral development projects without Indian input and consent.

Debt and Resources

Bringing some of these ideas to fruition will require cash, and the indigenous people for the most part have none. "Debt-for-nature swaps" are one way to assist indirectly in bankrolling some of the desired changes. First proposed by ecologist Thomas Lovejoy in 1984, the debt-for-nature swap is simply a new application of the debt-for-equity swap concept and could involve, for example, a timber company paying off some government debt in exchange for logging rights.

Though these debt-easing deals sound good, and in some cases work well, they are no cure-all for the problems confronting the rain forest and its inhabitants. For one thing, very little debt is for sale. And debt that can be renegotiated often has dubious value. The reduction saves a country money it didn't really have in the first place. As a result, Barbara Bramble of the National Wildlife Federation believes that "putting a million dollars into conservation will often take a million away from another project such as child care."

Even to begin to understand the Indian battle for land titles in Peru—and, ultimately, rain forest preservation—one must look at the basic political context in that country. Peru is financially paralyzed by a staggering foreign debt. To meet its payments, it is doing what other countries in the Amazon region do that are in the same situation: exploit the vast resources of the rain forest.

In addition to the debt, or perhaps because of it, Peru suffers chronic inflation. For example, a U.S. dollar was worth 28,000 Intas the first day of the COICA conference, and 33,000 on the last. In other words, relative to the international economy, Peruvians lost one-fifth of their buying power in about a week. Of course, with an economy such as this one, nobody saves any local currency. That's why the central bank is broke; most people withdrew all their money a long time ago.

To complicate matters further, Peru grows about 80 percent of the South American coca leaf crop. As much as they'd like to, the owners, growers, and traders of this notorious cash crop don't operate in a vacuum. Intimate connections exist between the forces behind the crop production and the Shining Path guerrillas. The Shining Path gets a percentage of the crop profits, which helps keep them well-armed to continue fighting one of the world's bloodiest and least understood civil wars.

Pawns

In the scope of this and many other conflicts involving the Peruvian military, the Colombian drug cartel, the Shining Path, and the U.S. military groups, the Indians are barely pawns. It is a sad irony since the indigenous people comprise a majority of the Peruvian population.

The situation in Bolivia is no less complicated. Until 1986, the Chimanes forest, for example, in the Beni region of Bolivia, was an untouchable forest reserve of 1.2 million hectares (or almost three million acres). It wasn't accessible by any roads, and was home to thousands of indigenous people. Lumber companies faced with dwindling reserves in other parks began pushing the government to lift the ban on logging in the Chimanes. Twelve

"IT'S OUR RAIN FOREST"

"We, the indigenous peoples, have been an integral part of the Amazon biosphere for millennia. We use and care for the resources of that biosphere with respect, because it is our home, and because we know that our survival and that of our future generations depend on it. Our accumulated knowledge about the ecology of our forest home, our models for living within the Amazon biosphere, our reverence and respect for the tropical forest and its other inhabitants, both plant and animal, are the keys to guaranteeing the future of the Amazon Basin. A guarantee not only for our peoples, but also for all of humanity. Our experience, especially during the past hundred years, has taught us that when politicians and developers take charge of our Amazon, they are capable of destroying it because of their shortsightedness, their ignorance, and their greed."

Coordinating Body for Indigenous Peoples Organization, "It's Our Rain Forest," **Mother Jones,** April/May 1990

companies petitioned the government, and seven were awarded permanent concession on about 600,000 hectares. Construction of roads immediately followed.

Throughout this entire process, no one from the government or industry consulted with a single indigenous resident. Meanwhile, Conservation International was orchestrating the world's first debt-for-nature swap. The deal went down in early 1987, and though it looked good from the outside, it too suffered from the same oversight as all the other deals: the indigenous people, the local inhabitants who would be most affected by any change in the area, were never consulted. According to COICA, the deal took place "with the most brazen disregard for the rights of the indigenous inhabitants and resulted in the destruction of the very forests the swap was meant to preserve."

In 1987, Indians clashed repeatedly with loggers and ultimately brought their case before the Bolivian government. After much discussion, the government reversed itself and in 1988 the president issued a Supreme Decree: until more information was available concerning the impact of harvesting the Chimanes forest, all lumbering and building would stop.

The decree, however, was never honored and the logging

continued. Consequently, the indigenous people mobilized in an effort to fight what they saw as essentially a conspiracy between the government and the private development organizations.

The Indians claimed that the government, when dividing up the Chimanes forest between industry and locals, gave all the good land to the timber concessions, and all the swamps and unarable portions to the Indians. Furthermore, the land that was given to the Indians was parceled into little pieces, effectively dividing their community and usurping their power.

The Chimanes Indians are maintaining their position in the ongoing struggle: They want the whole forest back. In addition, they want each family to hold title to 200 acres, instead of the 20 acres currently allocated. Even if their demands are met, and the forest is returned to them, the Indians face a formidable host of tough problems. Their numbers are growing, with no real discussion of population control. And resource management will be an issue, as they too wish to do some logging.

The Future

The rain forest is a big place that is shrinking fast. There is not a lot of time left to figure out what to do to help, given the current rates of destruction. (The World Resource Institute recently published findings that we were wrong in our estimates of the rain forest disappearing at a rate of one acre every second; it's actually going at one-and-a-half acres every second.)

The Iquitos Summit was a positive step. But many tough decisions remain. The development of a policy and a plan that will protect the rain forest requires balancing an enormous number of facets and interests — a balance upon which the Indians, the forest, and ultimately our entire planet depend.

As COICA president Evaristo Nkuang of Peru said, "We either disappear with the forest, or live with the forest. We have no other place to go. And we need a place to give our children."

AT HOME IN THE FOREST: BRAZIL'S KAYAPO INDIANS

Douglas H. Chadwick

Douglas H. Chadwick is a field correspondent for Defenders of Wildlife, *an environmental organization dedicated to the preservation of all forms of wildlife.*

Points to Consider:

1. How primitive are the farming methods of the Kayapo? Explain.

2. How do the Kayapo people use plants as medicine?

3. Why are the burning methods of the Kayapo less harmful to the rain forest?

4. What lessons can be learned from the Kayapo? Give examples.

Douglas H. Chadwick, "Home in the Amazon", **Defenders,** July/August 1988.

The Indian agricultural practices integrated with these forests are extremely sophisticated by any standard.

Lands drained by the Xingu River, a major tributary of the Amazon, lie mainly in the state of Para, at the geographic center of Brazil. Among the array of Indian tribes dwelling within its forest are the Me be nqo kre—"people of the waterfall". That is what they call themselves, being linked to waterfalls in their origin myths. Others know them as the Kayapo. Several rather large villages can be found within their territory. On a bank of the Rio Fresco in the Xingu's headwaters, stand the mud and thatch houses of Gorotire, which I reached toward the end of the dry season in July.

I was introduced to the natives of Gorotire by a blond, blue-eyed native of Kentucky, Darrell Posey. Darrell's Kayapo name—his "beautiful" name, as the Indians put it—is Tegmai, meaning "one who digs roots". They also call him Totn'i, after a famous warrior, and Yarati, which simply means "big white person with a beard".

This ethnobiologist—ethnoentomologist, to be precise—has worked with the Kayapo for a dozen years. He is overseeing a multidisciplinary investigation of virtually everything in this tribe's sphere, from kinship patterns to bird densities to shamanistic visions. The real purpose is not so much to study the Kayapo as to learn how the Kayapo themselves study their surroundings.

Advanced Farming Techniques

The Kayapo take a great variety of fish from the forest waterways and add other meat when the opportunity presents itself: monkeys, deer, armadillos, tapirs, white-lipped and collared peccaries, large rodents such as pacas and agoutis, land turtles, river turtles and their eggs, lizards and an assortment of large birds. However, the bulk of their diet is vegetarian. And, importantly, they do not really forage for wild plants. Rather, they farm plants in the wild.

As the dry season ends, the season of planting comes to hand. Nature announces the best time to sow by the position of the Pleiades constellation among the stars, the purple flowering of a treetop vine, the arrival of kestrels, the mating call of a certain cicada that, to me, rings like pure crystal, and the shape of cloud formations. All of these herald the rains that begin as afternoon showers and increase to day-long downpours.

A family of Caboclos, or people of mixed blood, deep in the heart of the Amazon rain forest. Source: U.N. Food and Agriculture Organization.

Seeds and plantlings will be placed along jungle paths, at distant campsites and in "resource islands," which are patches of useful plants cultivated in drier *cerrado*, or upland savanna habitats. People plant trail edges casually, taking seeds from something they happen to be eating en route and tucking them into the ground. They also plant purposefully and with ceremony, as when children go out along the same trails to seed trees that won't bear fruit until their own children's or grandchildren's time. In addition, the Kayapo develop intensively managed gardens on cleared and burned plots within the jungle.

Having seen far too many pictures of tropical rain forests left the way I saw them from the airplane window—scorched and bare—we have come to view slash-and-burn agriculture as a terribly destructive force. On a large scale, it is indeed a disaster. Little is produced over the long run but impoverished soils and impoverished colonists. Yet on a small, carefully managed scale, as practiced by indigenous people, the slash-and-burn method can prove very fertile.

Practically all the available carbon and nitrogen in a tropical rain forest end up stored in live vegetation. The surest way to release those vital organic elements is by burning. After a fire, you may see the new sprouts becoming thickest along a strip of ash left by an individual tree. Tilling only increases erosion and the leaching out of valuable minerals in an environment where rain falls as hard and heavy as it does here.

Susanna Hecht suggests that in the Amazon forest, it is these farming and grazing techniques of the white man — not Indian slash-and-burn cultivation methods — that are "primitive".

The Kayapo distinguish among 30 or 40 different types of forest here — far more than ecologists were aware of until recently, and certainly far more than colonists understand. The Indian agricultural practices integrated with these forests are extremely sophisticated by any standard. Covering 8.1 million acres, Kayapo territory is roughly the size of France, and the traditional trading network with other tribes draws resources and information from an area the size of western Europe. Thus many crops raised here originated far afield. Others, such as *kupa*, the starchy vine that I saw trained up tree trunks at the edge of gardens, were domesticated locally.

Besides accumulating a healthy variety of crops, this tribe has developed a variety of genetic strains within each staple, making it less vulnerable to being wiped out by disease or some change in the environment. Some 22 different cultivars, or breeds, of sweet potato, 22 or more of manioc, 21 of yam and 21 of corn go into Kayapo gardens.

A number of foods — especially in the resource islands — are grown not only for human consumption but also to draw in animals which can then be hunted more easily. Using a similar principle, the Kayapo cultivate "trap" plants around their garden plots to attract insects, diverting them from the crops.

Modern Medicine

Medicinal plants are sown in convenient sites around villages and in most other places where crops are planted. Hiking down a trail with Ute, a village chief, I at first saw only what looked like rampant jungle growth on either side — a barrier to be hacked through with the machete. Through his trained eye, though, it reappeared as a continuous, mostly man-made cafeteria and pharmacy, not to mention a warehouse of fabric and building materials. Here was a small edible *bromeliad,* or pineapple. Across the way stood a yam plant, carefully bred from wild strains. Shortly, we stopped to examine a tree whose bark is said to cure rashes, then a shrub used as medicine for dogs. The leaves of a vine snaking between them could be rubbed over the body of an electric eel victim to help overcome the effects of shock.

So attuned are the Kayapo to their surroundings that they can literally see beelines—the flight paths steered by the Amazon's tiny stingless bees through dense vegetation.

According to Darrell, some 1,200 plants have been collected by the research team to date. Of those, 98 percent have some specific use among the Kayapo. Darrell is betting that the figure will be 100 percent when he and his colleagues learn more from the natives. About 45 percent of the plants have proved new to science. Some 90 percent of the uses were also new to science. The other figure to note is that 95 percent of Amazonia's plants—and insects, along with other animals—will probably be extinct before anyone has analyzed their active chemical compounds or other properties. What protein sources, what fertilizers, what future materials, what near-miraculous cures, what possibilities of a better, fuller life are being cast away?

Beptopup, a shaman, specializes in two types of medicinal plants: first, those that cure snakebite and scorpion stings; second, those that increase spiritual energy. He knows how to splice his medicinal plants onto fruit trees to keep them available near the village, much as other shamans graft orchids with contraceptive properties onto handy trees.

Still Learning

As for the Kayapo, they are still learning—still trying to pin down what grows where and how it is affected by sunlight or shade, leaf-cutting ants, "magic" forces or specific soil types. Their detailed system of soil classification is based upon many of the same factors that any other kind of soil specialists would consider: sand content, clayiness, color and organic inclusions. A surprising ten percent of Amazonia is black earth, or anthropogenic soil, unusually rich and fertile. In other words, it is manmade dirt—the consequence of fire and agriculture by native people, practiced over millennia.

"In the case of the Kayapo," Darrell continues, "consider that before the tribe was forced into contact with whites, villages might number five to six thousand—ten times what they are today. Imagine the influence those people were having on the nearby environment. And on more distant parts of the forest. Groups of up to 160 hunters would go out for up to four weeks at a time, covering perhaps 480 square miles, changing campsites every two nights, planting seeds. They were altering habitats wherever they went."

Darrell believes that the supposedly natural distribution of many prized Amazonian foods such as the Brazil nut is really, like black earth, a human artifact. For how many tropical species used as medicine or fiber is this also the case? Certainly the common image of native tribesmen trapped at some subsistence level of foraging needs a lot of revision. There is little question but that they have shaped a broad array of biotic resources through their agricultural practices and genetic selection.

Preserving Knowledge and Culture

The question now is what will become of that native knowledge? The Kayapo nation has been reduced from at least 30,000 members at the turn of the century to around 3,000 today. The Gorotire Kayapo reached a low point in 1945, when the village population sank to 85. White man's diseases had taken a terrible toll. So had intertribal fighting as whites displaced Indians from traditional territories. Moreover, many adults—perhaps discouraged by a powerful shaman's vision of a cattle ranch on top of their village—simply did not want to have children.

Today the Gorotire population is roughly 720 and increasing, as are most other Kayapo groups. However, new troubles are headed their way. Commercial farming and cattle-raising in the rain forest have proved far less profitable than hoped. Yet enough gold and diamonds keep appearing in the Amazon basin to trigger new dreams of easy wealth, along with further colonization. Miners, or *garimpieros*, already have leased portions of the Kayapo reserve within 50 miles of Gorotire for mining operations. The tribe receives royalties but has paid dearly for them. Where the Rio Fresco used to flow clean and bright past the village, it now runs yellowish brown with mud churned up by mining along a tributary. Gorotire's people are compelled to paddle a couple of miles upstream for clear current and the rainbows of fish I admired.

Logging companies, too, have leased some land from the Kayapo, resulting in a series of roads being pushed into the area in 1985 from an already existing branch of the Trans-Amazonia highway.

As each stretch of forest disappears, so do a number of young Kayapo and the knowledge they inherited. They drift away into the lumber camps, gold mines and frontier shantytowns, becoming more nameless faces among the burgeoning mass of people in a debt-ridden Third World economy. I find the loss especially poignant because traditional Kayapo society could hardly be more different from the scramble for gold, money and material goods closing in upon it. Chiefs would typically own the poorest houses in the village, having given away nearly all that they had. Their power was based upon that generosity—upon how many favors they could call in when needed.

In the end, what we're talking about saving, or losing, is knowledge. Knowledge—in the form of plants and animals with all their adaptations—about how best to live in a certain environment. Knowledge—in the form of indigenous people with generation upon generation of accumulated learning—about how best to use living resources. If governments planning the future of the Amazon and similar regions can recognize this, perhaps we may begin to redefine progress. Progress is the exact opposite of destroying nature. Progress is moving from knowledge toward wisdom.

8   GENOCIDE AND NATIVE PEOPLE

THE ONGE:
A MATTER OF SURVIVAL

D. Venkatesan

D. Venkatesan has done extensive anthropological fieldwork among the tribals and nontribals in the Andaman and Nicobar Islands of India from 1984 to 1987. He works for the government of India's Anthropological Survey. This reading deals with the Onge, who inhabit India's Andaman Islands south of the mainland in the Indian Ocean.

Points to Consider:

1. Who are the Onge and where do they live?

2. Why are they being resettled?

3. How are the rain forests being exploited in the Andaman Islands?

4. How are the Onge threatened with extinction?
    Give examples.

D. Venkatesan, "Ecocide or Genocide? The Onge in the Andaman Islands", **Cultural Survival Quarterly,** vol. 14(4). Reprinted with permission of **Cultural Survival,** 53-A Church Street, Cambridge, MA 02138, USA.

Resettlement has set in motion the biological, social, and cultural death of the Onge.

British colonial expansion profoundly altered indigenous people's demographic structure, subsistence mode, and social organization not only in the Andaman archipelago, but around the world. As colonialism forced indigenous people to transform from foraging to advanced hunting and gathering, to shifting or settled agriculture, some of the foraging communities completely lost their will to survive and now approach extinction. One such foraging society is the Onge, a group of about 100, one of the four hunter-gatherer groups in the Andaman Islands of India.

Onge contact with the British began back in 1825, and has claimed many Onge lives. Finally, after continuous efforts and attempts to control the island, the British established "friendly" relations with the Onge in 1883. The British took interest in these islands for their rich rain forest resources, especially the timbers. This policy of exploitation continues today, as the government of India continues to extract forest resources. Deforestation poses many problems to the Onge in terms of demography, subsistence, ecology, and culture. They are a group that has undergone dramatic change, not only since their initial contact with "outsiders", but particularly after they were "resettled" in 1976.

Resettlement of the Onge

In the 1950s and early 1960s, while Little Andaman Island was inhabited by the native Onge and temporary residents (mostly government employees, numbering about 1,500), the government of India passed a resolution to resettle the refugee population from East Pakistan (now Bangladesh) to Little Andaman. Up until 1961, 3,281 refugee families received four hectares (10 acres) of forest land each. Between 1964 and 1973, Andaman administration cleared 51,400 hectares (128,500 acres) of lush green forests, out of which about 10,000 hectares was claimed in Little Andaman for settlement and cultivation.

The Onge on Little Andaman were the most affected ecologically during this rehabilitation process. The island's flat land made it ideal for 2,200 refugee and repatriate families. Almost 9,000 hectares (22,500 acres) of coastal forest was cleared before 1969, and another 20,000 hectares (50,000 acres) was proposed for clearance in the early 1970s on the northern part of the settlers' village, near Onge territory.

Different development schemes in Little Andaman, including

Logging operations on Sri Lanka. Source: U.N. Food and Agriculture Organization.

allocating land to settlers, planting monocrop species, constructing roads, and building government offices, private industries, a harbor, a subnaval base, an agriculture farm, and a helipad, have all added to the large-scale forest clearance. The Onge have been forced to move away from the outsiders' settlements into smaller pockets in the southern and northern parts of the island. In 1976 the Andaman administration set out on a deliberate scheme to "resettle" the Onge to an area in the island's northeastern corner known as Dugong Creek, providing them with mainstream amenities and coconut plantations.

This "resettlement" has set in motion the biological, social, and cultural death of the Onge. Their natural forest resources have been reduced from 292 sq. mi. to 44 sq. mi., creating a crisis in their subsistence pattern. Moreover, the government has engaged them as wage laborers on the coconut plantation, further hindering their traditional subsistence activities.

Deforestation

Exploiting rain forests in Andaman for timber and rehabilitating uprooted people often means encroaching onto the traditional resource base of the hunter-gatherer communities such as the Onge. The implications of clearing and extracting timber from these islands in general and from Little Andaman in particular are not considered seriously enough by the government, either in terms of ecology or human suffering. Both public and private agencies are engaged in timber extraction, and commercial exploitation of timber species has shot up from four in 1952 to forty species today. Ninety-three wood-based industries are now operating in Andaman.

During 1984-1985, the timber extraction in these islands was 133,942 cubic meters, and the revenue earned by the Forest Department was 76 million rupees (U.S. $5.7 million), the highest income ever made in a year. At present, the timber extraction has stretched to 150,000 cubic meters. The major share contributed to this income was by Little Andaman forests losing about 20,000 cubic meters of timber every year.

In addition to the large-scale timber extraction and clearance of Little Andaman forests by government and other agencies, the settlers and the recent migrants, although they are allocated sufficient land, have begun illegally encroaching on the adjacent forest land because of Free Royalty on timber use and demand on land area. Unless the over-felling and illegal destruction of forests ceases in Little Andaman, the Onge, who depend on this natural resource for nutrition, will face acute food shortages.

Resource Depletion

Little Andaman forests once were abundant with wild boars, roots, tubers, fruits, and honey, and the sea with fish, turtles, and dugongs. Depending on the season, the Onge have adapted their diet according to availability. The depletion and degradation of forests, however, has led to a sharp decline in the Onge's natural food resources, which in turn has led to malnutrition. Settlers—particularly the Bengali settlers—are increasingly hunting wild boars and other forest and sea resources.

As a result, the Onge have to spend more hours covering a greater distance to hunt and gather; most of the time the men return without game. They have had to increase their dependence on marine resources, which once merely supplemented their land-based foods. The Onge are now entirely dependent on government doles, and their once primary hunter-gatherer subsistence has become secondary.

Population Decline

Many factors, related to the loss of their traditional resource base and the increase of outsiders, have led to the Onge's population decline. At present there are about 12,000 outsider settlers, both temporary and permanent, in Little Andaman—a population density of 16 people per square kilometer, as compared to two people during the 1950s and early 1960s.

Various anthropologists studying Onge population decline in the last few decades have emphasized the diseases introduced by outsiders. The high levels of sterility among Onge women and of infant mortality in the last two decades reveal a bleak future for the Onge.

Due to the resettlement resulting from deforestation, the Onge have taken on many of the cultural traits of outsiders, including drinking alcohol, which was recently introduced by two Onge boys who were recruited into the local police force. The food supplied to the Onge plays a significant role in intestinal diseases, diarrhea, and dysentery. The Onge are now motivated by a cash economy, which can provide only short-term satisfaction.

A Matter of Survival

Various developmental programs implemented among the Onge have failed drastically, no doubt due to ethnocentrism. Replacing the once-abundant Onge traditional food with modern food is not going to provide sufficient nutritional balance, since the Onge have always depended and survived on natural nutrient energy intake. Their protein intake has been reduced enormously. One acre of forest land has been turned into a vegetable garden, but the Onge never utilize the yield; instead, it fulfills the needs of the non-Onge families in the settlement. Intensive horticulture could be initiated in the Onge settlement, giving more emphasis to vegetable growth to counter the protein deficiency.

Although the Onge fish daily in the shallow ocean waters near the settlement, their catch is not encouraging. Between October and December the yield is quite reasonable; however, the marine resources are being exploited by those settlers who use advanced fishing techniques.

Extending a road to Dugong Creek; settlers exploiting wild boars, honey, fish, and turtles in the "Onge Reserve"; proposed forest clearance of 20,000 hectares (50,000 acres) toward Onge territory for non-Onge settlers; keeping the Onge as "museum pieces" for VIPs' visits; exploiting the Onge economically and encouraging them to involve themselves in more cash income based on illegal dealings—all these activities should cease immediately. Unless the administrators, planners, and social scientists take the current situation seriously, the Onge will face a crisis of survival in a very few years.

DETECTING CULTURAL BIAS AND ETHNOCENTRISM

This activity may be used as an individualized study guide for students in libraries and resource centers or as a discussion catalyst in small group and classroom discussions.

*Good readers are aware that written material usually expresses an opinion or bias. The capacity to recognize an author's point of view is an essential reading skill. The skill to read with insight and understanding involves the ability to detect different kinds of opinions or bias. **Sex bias, race bias, ethnocentric bias, political bias and religious bias** are five basic kinds of opinions expressed in editorials and all literature that attempts to persuade. They are briefly defined in the glossary below. This activity will focus on ethnocentric bias.*

Five Kinds of Editorial Opinion or Bias

Sex Bias—the expression of dislike for and/or feeling of superiority over the opposite sex or a particular sexual minority

Race Bias—the expression of dislike for and/or feeling of superiority over a racial group

Ethnocentric Bias—the expression of a belief that one's own group, race, religion, culture or nation is superior. Ethnocentric persons judge others by their own standards and values.

Political Bias—the expression of political opinions and attitudes about domestic or foreign affairs

Religious Bias—the expression of a religious belief or attitude

Guidelines

Read the statements below and then answer the questions that follow.

Statements

A. Many people from the rich nations are horrified at the destruction of the planet's tropical rain forests. Issues of concern include loss of wildlife, global warming and climate changes, ozone and oxygen depletion, and the displacement or killing of native Indian people. Something must be done soon to force the countries involved to slow or halt this trend. The developed nations should lead the way and promote establishment of massive national and international forest preserves; encourage governments to halt further construction of roads and dams in the Amazon; make monetary loans only on the condition that steps be taken to save the rain forests; and establish an international agency to monitor these actions.

B. The eastern half of the United States was once covered by mighty forests stretching from the Atlantic coast to the Great Plains. By the end of the last century, most of these forests had been cleared for roads, farms and timber use as the U.S. grew into an agricultural-industrial giant. Today, there is little left of this vast forest. Logging continues in the east and threatens the only remaining stands of virgin growth timber in the Pacific Northwest and Alaska. Deforestation in the U.S. continues to cause soil erosion, stream pollution, and floods while habitat depletion threatens endangered wildlife.

Questions

1. Do the above statements illustrate a case of cultural bias?

2. Compare and discuss the treatment of native groups in the Amazon today and the Indians of North America during the last century.

3. Read each statement below. Indicate with an **(X)** the statements that illustrate a cultural or ethnocentric bias. Use an **(O)** for those statements that do not.

   ____Deforestation threatens human as well as plant and animal life.

   ____The Indians of the Amazon should learn modern farming techniques to manage the rain forest.

_____Scientists should continue to study the biodiversity of the world's rain forests.

_____The nations of Central America should establish more national and international parks to preserve their forests.

_____Miners and homesteaders should not be allowed to enter lands inhabited by indigenous people.

_____At the present rate of deforestation, up to one-fifth of the world's plant and animal species could disappear by the year 2000.

_____Pharmaceutical firms have the legal right to patent their discovery of medicines used by native people in the forests

4. Can you locate any cases of cultural bias from the readings in this publication?

CHAPTER 3

SAVING THE FORESTS:
IDEAS IN CONFLICT

9   SAVING THE FOREST: IDEAS IN CONFLICT

"SAVE THE FOREST" HYPE IS JUST A FAD

Jon Christensen

Jon Christensen wrote this article for Pacific News Service *after a one-year assignment in Brazil.*

Points to Consider:

1. What is the "fad" discussed by the author?

2. Why does he believe many concerns over the rain forest are unwarranted?

3. How does this article contrast with other readings in this publication?

4. How are Amazonians working to solve their problems?

Jon Christensen, "Our 'Save-the-Rain-Forest' Fad, Just Another Jungle Fantasy," **National Catholic Reporter,** January 26, 1990.

Experts disagree about the extent of deforestation—but it is clear that the rain forest will not disappear any time soon.

America's obsession with the Amazon reveals more about ourselves than it does about the rain forest.

The current "save-the-rain-forest" fad in the post-industrial countries of the world is not much different from earlier jungle fantasies; in both cases, the jungle played a role as our opposite. When we believed in science and progress, the jungle was a primitive, bewildering, even terrifying green hell. But the greenhouse summer of 1989 triggered a growing fear that humanity will not survive without the mysterious forces embodied in the last of the great wild lands.

Aura of Romanticism

Because we are worried about suffocating from progress, the rain forest has become a repository for romantic notions about primitive life—a Garden of Eden filled with miraculous plants and animals and Indians who live in pre-industrial harmony with nature.

An aura of romanticism and First World self-righteousness has permeated Sting's "Virgin Forest" campaign, Madonna's "Don't Bungle the Jungle" concert and other pop efforts to save the planet by saving the rain forest. Simple messages have won out over a complex reality. The fact that 15 million people now live in the Brazilian Amazon—more than half of them in cities—has been virtually ignored.

The Amazon does not serve as the lungs of the planet; the forest consumes as much oxygen as it produces. It is not being cleared to produce fast-food hamburgers; the region is a net importer of beef. Experts disagree about the extent of deforestation (reasonable estimates vary between eight and twelve percent, mostly along the edges of the jungle); but it is clear that the rain forest will not disappear any time soon.

Scientists are also uncertain of the role of rain forests in the globe's climate. And while Amazon burning adds to the greenhouse effect, the fires annually contribute less carbon to the atmosphere than exhaust from cars in the United States.

Slash and burn clear-cutting in Northern Brazil. Source: U.N. Food and Agriculture Organization

Villains and Saviors

In the rhetorical firestorm engulfing the Amazon, people from distant First World cities have eagerly come up with lists of villains and saviors in the rain forest saga, but they seem less interested in what is really happening there now. Many critics consider fire the number one enemy of the forest. But agriculture in the Amazon depends on burning to prepare poor soils for planting, whether the farmer is an Indian, a poor migrant or a wealthy rancher.

Outside the region, the chainsaw is also considered an enemy of the forest. There has been a suggestion that Amazonians be required to register their chainsaws as if they were weapons. That might seem a good idea from the glib distance of Los Angeles, London or even Rio de Janeiro. But in actuality, it would be a bureaucratic nightmare, aimed at what is, in fact, a labor-saving tool for forest dwellers around the world.

Roads are another favorite target. After conducting a "fact-finding" tour, U.S. Senators John Heinz and Timothy Wirth fanned environmental fires at home by declaring that they had seen heavy machinery poised to rip a road through the jungles from Brazil to Peru. About that same time, I sat perplexed at the end of BR-364, the "highway" in question. There hadn't been any road-building equipment in the vicinity for years. Residents of the remote northwestern Amazon have been waiting decades for the

MARKETING THE RAIN FOREST

Clearly, the best way to protect the viability of forest communities and to ensure the future of their resources is to expand the market for rain forest products. At Cultural Survival, *we have learned over the years that without income, indigenous people have little chance of defending themselves. Over the past few years we have funded sustainable resource management projects with indigenous peoples who produce commodities that would benefit from expanded markets. From this experience, we have established Cultural Survival Imports, a nonprofit trading company.*

We are currently collecting a wide variety of product samples, including nuts, fruits, oils, flour, fragrances, tubers, spices, and fiber and medicinal plants. The most likely new markets for fruits, for instance, are in products that could be flavored with, rather than made entirely from, an imported fruit: ice cream, yogurt, milk, candy, soda, snacks, wine coolers, and various natural food products.

Leslie Baker, "Cultural Survival Imports: Marketing the Rain Forest," **Cultural Survival Quarterly**, No. 13, Vol. 3, 1989

government to pave the muddy track that connects them with the rest of Brazil for only six dry weeks out of the year.

It is true that chainsaws and fires have cleared the way for enormous, unproductive cattle ranches and land speculation. Roads facilitate what has often turned into uncontrollable immigration and failed settlement projects. And dams that are too big to work well have forced Indians and other forest dwellers to relocate in order to provide electricity for industries and cities, which some might consider an alien presence in the rain forest. But we could not turn back the clock on the Amazon even if Brazilians wanted to do so—and they don't.

Solving Their Own Problems

Meanwhile, how do rain forest activists propose to "save the Amazon"? Much has been made of using the ecological wisdom of native Amazon Indians to exploit the resources of the rain forest without destroying them. It is an article of faith that indigenous knowledge somehow will provide an answer for the future. But no one knows whether the intricate forest management practices of the Indians—sharpened over many generations in particular ecosystems and usually encoded in story and myth—are applicable on a large scale. The only Indian technique that has been successfully adapted so far is clearing the forest with fire.

There is still far too much to be learned about tropical forestry, fisheries and farming to believe we now know what is best for the future of the Amazon. The region is as vast as the American West, encompassing myriad resources and ecosystems, making it foolish to try to impose blanket solutions to the practical problems of the people who must make a living in order to survive there.

Amazonians, sometimes with the quiet help of foreigners, are working on solving their problems in small ways and in particular places. The Brazilian government has cracked down on illegal burnings and suspended development subsidies that encouraged indiscriminate clearing. Amazon states have drawn agro-ecological zoning maps and adopted plans for development appropriate to local soils and forest cover. And now under Brazil's new constitution, the Indians of the Amazon are increasingly representing themselves.

In the far northwestern state of Acre, rubber tappers are working with the local government to diversify the crops they harvest and find new markets for their produce. Foresters are experimenting with selective logging and reforestation. Cities are trying to attract industries to bring jobs to the region and add value to exports instead of shipping out low-return raw materials.

But "save-the-rain-forest" hype is obscuring the need for painstaking, on-the-ground work. Our new romance with the rain forest may satisfy a need to feel we can play a role in the fate of a last great wilderness, but it will not go very far in helping the Amazon and its people survive into the 21st century.

10   SAVING THE FOREST:
IDEAS IN CONFLICT

SAVING THE FOREST IS A
REAL CONCERN

James P. Doyle

James P. Doyle is a student at Front Range Community College, Westminster, Colorado.

Points to Consider

1. How fast are the rain forests disappearing?

2. Who are "the Lost Ones"?

3. How is Brazil's debt to the World Bank responsible for the destruction of the rain forests?

4. What effect will iron ore mining have on Brazil's forests?

James P. Doyle, "Rain Forests Are Battlefields of Our Survival", **National Catholic Reporter,** December 8, 1989. Reprinted by permission, **National Catholic Reporter,** P.O. Box 419281, Kansas City, MO. 64141.

The scale of destruction in Brazil is severe enough for some analysts to predict the complete elimination of the rain forest by the end of this century.

In a broad sweeping arc it fell. Two hundred feet in vertical growth plunged to a splintered end. The twisted branches and leaves shuddered violently on impact. Another green domino fallen in a wave of destruction called progress. Every minute of every day finds 30 more acres of tropical rain forest leveled in a ribbon of deforestation.

Cutting hour after hour, the chainsaws will send four more monoliths crashing through the green canopy of the jungle before noon. The mahogany and teak, the cherry and liana fall and wait. They wait for the sun to dry them. They wait for the gray curtain of smoke to envelop them. They wait for the coming inferno to consume them, forever gone.

Only a few thousand years ago, rain forest carpeted most of the equatorial regions of the earth. In the last two decades, hundreds of thousands of acres of pristine forest have been converted into a wasteland capable of supporting only a few tough, fire-resistant weeds and cattle. These methane dispensers feed not only Brazil, but also the world's feverish appetite for fast food.

Many Casualties

The trees and plant life are not the only casualties of modern humanity's assault on the Amazon Basin. This expanse of jungle has been home to many Indian tribes in the past 100,000 years. For one tribe in particular, the Ureu Wau Wau, the world outside of the Amazon did not exist until their discovery by field scientists in the 1960s. Alone and alive, these people have withstood the wrath of the ages, living in the forest and becoming a part of its delicate structure.

Known in Brazil as "the Lost Ones", the Ureu Wau Wau now fall with the timber, not to the ferocity of the chainsaws and torches, but to the disease and murder thrust on them by the presence of civilization.

As you read these words, several Brazilian Indian tribes face immediate extermination. According to anthropologist Berta Ribeiro, the Asurinis tribe has been destroyed by intruders and the diseases they pack in their belongings. Only 53 Asurini are left, Ribeiro states, and "these Indians want no more children; they know they are finished."

Domestic use of fuelwood in Senegal, Africa. Source: U.N. Food and
Agriculture Organization

Whole civilizations are being lost to a fever they cannot
control. Our lust for gold and our need for minerals have helped
incubate this holocaust.

Scale Is Severe

The scale of destruction in Brazil is severe enough for some
analysts to predict the complete elimination of the rain forest by
the end of this century. With half of the world's plant and animal
species concentrated on less than 10 percent of the earth's land
surface, the destruction of this fragile habitat has become a
subject of global concern.

What had begun as an economic development plan to free
Brazil from poverty and unemployment has now become an
economic nightmare. Brazil now restricts unlimited migration into
Rondonia. But for large tracts of rain forest, it is too late. Still,
the people flock to Rondonia as surely as the mosquitoes
emerge after the rain. The latest waves of settlers must make
use of the burnt, eroded, abandoned land littered behind the

original pioneers. Their only hope is that the cattle they bring to graze will survive.

The mosquitoes that had lived in the jungle canopy and fed on primates and birds now swarm to the humans and cattle, the only warm-blooded animals left. Just as the chainsaws level the trees, the wracking pattern of malarial chill and fever cuts down the pioneer immigrants. In 1987 alone, there were 240,000 cases of malaria in Rondonia, roughly 20 percent of the population.

The cavalier and unstructured manner ·in which money has been lent to Brazil must force lending nations of the world to assume some responsibility for what is happening there. Causing a country to ravage its natural resources to make interest payments on its foreign debt is as destructive to the rain forest as holding the chainsaw.

Financial Considerations

Is the situation in Rondonia beyond hope? Are we left to extinguish what little flame of life is left in the rain forests of Brazil? The answers range from the hopelessly naive to the senselessly pragmatic.

No solution to the loss of jungle habitat will ever be enacted without addressing the financial status of countries like Brazil. Government officials in Brazil now realize that the development of the Amazon Basin cannot possibly begin to reduce a foreign debt estimated at more than $100 billion. Even after extracting the mineral wealth of Rondonia, Brazil cannot make the interest payment due on that debt.

Currently, several environmental groups have consolidated their financial resources to try to buy back some of the debt Brazil owes the World Bank. This debt, already heavily discounted, is being assumed for pennies on the dollar. However well-intentioned these environmental groups are, it seems unlikely that the debt will ever be eradicated in this way.

According to Thomas Lovejoy, president of the World Wildlife Fund, Brazil is about to open what is thought to be the world's largest and richest deposits of iron ore deep inside the Amazon in a state neighboring Rondonia. Lovejoy insists the development of this acreage will literally mean the end of the Brazilian rain forest.

The cumbersome weight of its debt threatens the very existence of Brazil as a sovereign nation. Until recently, the government of Brazil, in its zeal for developing the rain forest, has shown little concern for the environment. Recent responses

TOURISM IN THE FOREST

Officials compare controlling tourism in the Amazon with trying to stop a gold rush. They note that a savvy tour operator can earn an average year's wages in a single day. And, usually, they add, at the expense of the environment.

A wide swath of litter forms a paper chase along the jungle highways, and pollution, from plastic containers to sanitary napkins, defaces the serene surface of nearly every lake in the central state of Mato Grosso.

There are no provisions for garbage collection or sewers. High water carries waste down the millions of miles of tributaries into the Amazon River, through which flows one-fifth of the world's fresh water.

Unregulated fishing is so extensive that many districts are reported nearly deplete of some species.

Ty Harrington, "Tourism Damages Amazon Region," **The Christian Science Monitor,** June 6, 1989

to the concerns of its lending partners and the worldwide outrage at what is being done to the earth's lungs have caused Brazil at least to appear concerned.

A United Nations study group listed the elimination of tropical rain forest habitat as the major environmental concern of the UN in this decade. If the rain forests are eliminated in this century, as some scientists believe, the balance of animal species will not be far behind. The "Lost Ones" are all but gone now. The rain forests of Brazil look like smoldering battlefields. Our responsibility is not to fix blame, but to insure that humankind will stop the destruction of this planet.

11

SAVING THE FOREST:
IDEAS IN CONFLICT

BRAZIL'S GOVERNMENT SAFEGUARDS THE RAIN FOREST

The Brazilian Embassy

The Brazilian Embassy in Washington, D.C., prepared this report on Brazil's Pilot Program of development and environmental safeguards in the Amazon region.

Points to Consider:

1. Briefly describe the Brazilian government's "Pilot Program". What are its objectives?

2. What is sustainable development?

3. How has the government been successful in easing the pressures on the rain forests and its people?

4. What is the government's outlook for the rain forest?

Excerpted from a report on the Amazon region issued by the Brazilian Embassy office in Washington, D.C., Autumn 1991.

Over the last few years, the [Brazilian] government has substantially improved environmental policy and legislation.

The objective of the Pilot Program described below is to maximize the environmental benefits of Brazil's tropical rain forest consistent with its development goals. It is proposed that [international] donors provide financial resources in the form of grants or loans on a highly concessional basis for a major program prepared by the Government of Brazil, with the support of the World Bank and EC (Commission of European Communities), aimed at sustainable development in the Brazilian Amazon and other areas of tropical rain forest in Brazil.

Efforts at environmental protection will only be successful if they are compatible with the development goals of the Brazilian people and with the attainment of a reasonable standard of living for the residents of the region. Without the support of the local residents, preservation cannot succeed. The proposal, therefore, embodies efforts to test and develop sustainable economic activities in the Amazon environment.

Sustainable Development (*Economic growth that does not harm the environment*)

The recent development of the Amazon region is a clear indication that environmental protection and economic development must be tackled together. People in the region will push for improvement in their living conditions and for increasing access to the standards of living of a modern industrial society. In the absence of viable sustainable development options, predatory action on the environment will be difficult to curb, despite efforts by the government and environmental groups. The real challenge of a pilot program for the protection of tropical forests is to produce a sustainable development approach to the region that will guarantee economic progress together with environmental protection.

Mining is the growth industry of the Amazon. The mineral wealth of the Brazilian Amazon has been estimated at U.S.$3 trillion, with deposits of gold, tin, copper, bauxite, uranium, potassium, rare earths, niobium, sulphur, manganese, schist, diamonds and other precious stones, and possibly petroleum. New mineral deposits are still being discovered. In 1990, for example, new tin ore and gold deposits were found in Rondonia.

In spite of its immense size, the Amazon region produces a

Brazilian Amazon forest north of Manaus in Amazonia.
Source: United Nations Food and Agricultural Organization

relatively low amount of agricultural products. A number of crops have been grown successfully, including rice, manioc, sweet potatoes and perennial crops such as coffee, coca, black pepper, guarana (a fruit used for making beverages), juta (fiber) and palm oil. However, there are many obstacles to growing and marketing crops, including poor topsoil in the majority of the region, long distances and a deficient transportation system.

The timber industry is expanding rapidly in the Amazon, particularly the tropical hardwoods sought for exports. Although the export volume has increased considerably from 1980 onward, the value of exports has not grown proportionately. The exploitation of wood is highly selective. In 1978, 34 species were exported; five of which represented 90 percent of the total export volume. Mahogany accounted for 84 percent of the total timber exports to the U.S. in 1978.

Government Progress

In order to protect Brazil's biological resources, federal and state governments established a National System of Conservation Units. The legislation distinguishes a variety of categories of conservation units, by degrees of restrictiveness of exploitation of natural resources.

The Ministry of Agriculture operates several research stations in the region, oriented toward improving agricultural productivity to minimize negative environmental impacts of agricultural production. Over the last few years, the federal government has substantially improved environmental policy and legislation. The new Federal constitution, promulgated in 1988, included a whole chapter on environmental matters and designated several ecosystems as areas of national heritage, thus to be preserved. The Program Nossa Natureza, launched also in October 1988, provided the framework for rediscussing the environmental policy and for restructuring official agencies in charge of different aspects of environmental protection. Fiscal incentives and official credits to agricultural and cattle raising projects were suspended.

In 1990, the new government created the Secretaria do Meio Ambiente (SEMAM), directly linked to the Presidency of the Republic, as a regulatory and policy making body. As head of this new agency, a prominent environmentalist was appointed.

The National Environmental Policy

The basic objectives of the National Environmental Policy are:

- the adoption of the concept of sustainable development to replace the concept of development at any cost which prevailed until the late 1980s. That means the sustainable use of natural resources while ensuring higher levels and better distribution of income to the present generation. This objective involves the suppression of fiscal incentives to programs and projects, either public or private, entailing environmental degradation;
- the increased environmental awareness by means of formal and non-formal environmental education programs at all levels;
- the preservation of the main ecosystems;
- the implementation of ecological-economic zoning, as an instrument to define development programs and regional and sectorial planning, and,
- the integrity of indigenous communities and their culture.

Among the main objectives for the environmental policy put forward in the document are the sustainable exploitation of natural resources.

The main objectives for the social policy are the reduction of social tensions, inequalities and contradictions, the improvement in the pattern of distribution of benefits of development among the different social segments, and the preservation of the indigenous communities.

The measures of support to the indigenous communities would involve, among others, the acceleration of the demarcation of indigenous areas, the compilation and use of indigenous production techniques, the improvement of assistance services to be rendered according to the different cultural traits, and the adoption of measures to ensure the integrity of indigenous heritage.

The recommendations on mining include the promotion of research on exploitation techniques and production of equipment leading to the decrease of environmental damage. The recommendations on forestry include the establishment of rational criteria for approval and licensing of new forestry projects, and the improvement of the conditions of sustainable exploitation of non-wood products, among others.

In 1989, in order to prevent the recurrence of major burning in the Amazon that occurred in the 1987 dry season, the federal government launched a special emergency campaign to combat forest fires during the dry season in the region (June-September). The operation was carried out successfully with a modest financial support from the World Bank and a special budgetary authorization by Congress.

The outlook for the near and intermediate term is for the continuing decrease of deforestation rate due to the reduction of migration from the south. Other factors helping to save the forest include the suspension/modification of credit and fiscal incentives to extensive cattle raising and large agricultural projects, the improvement of legislation, law enforcement and control activities by federal and state agencies and, most of all, the increasing public concern with environmental preservation.

12   SAVING THE FOREST: IDEAS IN CONFLICT

GOVERNMENT AND GREED DESTROY THE FOREST

David Ransom

David Ransom is co-editor of The New Internationalist. *This article deals with the continuing deterioration of conditions in Brazil's Amazon, and the plight of its people.*

Points to Consider:

1. Why is poverty so widespread throughout the Amazon?

2. What was the objective of Brazil's military in the Amazon region? Explain.

3. How are roads affecting the rain forest?

4. How does the author view sustainable development?

David Ransom, "Gathering Strength", **The New Internationalist,** May 1991.

It was the military dictatorship that launched the modern invasion of the Amazon.

Look at the balance sheet. Clearing the Amazon produces hardwoods which are essential for nothing. Dams flood the forest to generate electricity to make aluminum for throw-away cans. Iron ore is dug out of the ground to be sold at throw-away prices. Diseased cattle stroll about farting their greenhouse gases into the atmosphere to pander to a particular dietary preference in the cities. Rivers are polluted with deadly mercury to produce gold that is smuggled out through Uruguay to languish in the vaults of Swiss banks. The mass of the people of the Amazon are corralled into poverty, fearing for their lives. Indigenous people are persecuted to the verge of extinction. Can it really be for this that the greatest forest on earth is being made to disappear?

Who Is Profiting

"Well," says Philip Fearnside, one of the top scientific authorities on the Amazon, who works in Manaus, "it's a question of who is profiting. If that profit—and the costs—were evenly distributed, deforestation wouldn't be happening. The fact is that influential people are making money and the poor people out in the forest are paying the cost. It's all perfectly logical from the point of view of the people who are making the money."

Anthony Anderson, another American expert, who works for the Ford Foundation in Rio de Janeiro, makes a similar point. "Ranchers can't have an environmental ethic", he says. "The key to environmental conservation in countries where the majority of people are poor is to engage them. They have a pretty innate sense that the forest is where their livelihoods remain."

But it is the ranchers and their like who rule Brazil. There are, it seems to me, just two simple reasons why they do. The first is that when real political change—and agrarian reform in particular—last threatened to become a serious possibility, there was a military coup. For 20 years after 1964 the country descended into a reign of official terror from which the people of Brazil have yet to recover fully.

It was the military dictatorship that launched the modern invasion of the Amazon. They built the roads and reinforced the network of cronyism on which the destroyers of the Amazon feed. To avoid political and social change they "colonized" the rain forest: "A land without people for people without land" was how the well-worn phrase went. In the process, oblivious of the

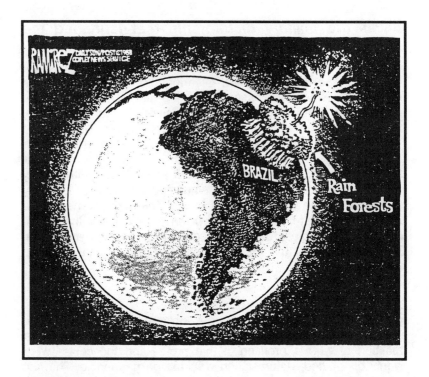

people who actually did live in the Amazon, and in pursuit of grandiose "strategic" objectives of the kind that appeal to the military mind, they saddled the people of Brazil with a foreign debt of quite astronomic proportions.

This brings me to the second reason why the ranchers and their like still rule Brazil. Quite simply, it suits their creditors. The military coup of 1964 was backed directly by the U.S. and tacitly by everyone else with a vested interest in the country. Today their priority is the repayment of the Brazilian foreign debt, not the preservation of the Amazon. The issue isn't that the destruction of the rain forest brings in lots of foreign currency with which to pay back the debt. It doesn't. More to the point is that real political change might lead to the renunciation of debt by people who feel, quite rightly, that they never signed the contract or benefited from its bounty.

Social "Safety Valve"

So the Amazon is made to serve as a social "safety valve", a sort of slush fund to keep the power brokers happy and the principle of debt repayment intact. What's so daft and almost pathologically criminal about it is that Brazil's creditors are in practice writing off the debt as fast as they can.

And, sadly enough, the international agencies that should know better are toeing the same line. Just about every expert on rain forests I talked to reserved his/her most virulent attacks for the United Nation's Food and Agricultural Organization (FAO). It is the "lead" organization in what is known as the Tropical Forestry Action Plan, which rides on the back of environmental concern while in practice promoting the interests of industrial forestry and forest-based industries.

It would be easy and possibly quite true to say that deforestation could be stopped tomorrow if: a) Brazil's foreign debt were written off, b) there were agrarian reform and c) cattle ranching, logging and mining were no longer promoted by the Government and international agencies alike. End of argument.

The trouble is that no one I met seriously believes that any of this is likely to happen tomorrow, or even the day after. And make no mistake, the situation of the people of the Amazon is deteriorating as fast as that of the rain forest itself—if not faster. So you have to try to puzzle out what might reasonably be done in the meantime. And that is precisely what many Brazilians have already been doing for some time.

I am told that there are upward of 2,000 "non-governmental organizations" (NGOs) at work in the Amazon. That might mean anything. But what is not really in doubt, I think, is that the unrest that began in the late 1970s marked a turning point not just for the military dictatorship but for many of the individual people and groups I spoke to as well.

Claudia Andrujar escaped from Hungary with her mother during the Second World War. The rest of her family did not survive. "I somehow always felt that I too was one of the people who had been condemned to disappear," she says.

She finally settled in Brazil and began a successful career as a professional photographer. By the mid-1970s she was beginning to tire of commercial photographic assignments. She took time out and went off to study the Yanomami Indians on the border between Brazil and Venezuela. She was to spend some six years living with them. "I saw disease arriving with the road. The people were totally unprepared. Then came the garimpeiros, the gold panners. I was there, witnessing the death of the people and trying to do something to help them survive."

Expelled from the area by the government, she launched a campaign in Sao Paulo to create a Yanomami Park. When I met her she had just returned from meeting the Minister of Justice,

MALARIA AND DDT

As if Brazil's Amazon jungle weren't already under enough pressure, a new World Bank loan for an "Amazon Basin Malaria Control Project" will add to its woes. The $99 million loan, matched by the Brazilian government, will provide, over a five-year period, a total of 3,000 tons of DDT to spread throughout the Amazon Basin for stopping the spread of malaria. In 1990, malaria infected more than a million people in Brazil. Although DDT has been used in the Amazon since at least 1983, the new loan expands the existing program to cover approximately two million square miles — almost the entire Amazon Basin.

Kay Treakle, "The Disease of Development: DDT Threatens Amazon," **Greenpeace,** November/December 1989

Jarbas Passarinho ("Little Bird"). He is a military man, the link between the civilian government of President Collor and the military establishment.

I asked him: "Well, what do you think? How are you going to help the Yanomami people?" He said: "You know, there's an awful lot of opposition to the creation of the Park." He talked a lot about "geopolitical considerations". I've heard this kind of talk since the 1970s and I thought they'd got over it. I thought President Collor wanted to change the image of Brazil. But at the moment he is not in a position to oppose the military.

Bringing Change

People like Claudia are clearly doing much to try and bring about change from within Brazil. But what can people outside the country do? "The most important thing you can do," says Claudia, "is to be willing to learn more and understand the problem, because it is only through understanding that you will find out what has to be done."

When I set out on this trip I was anxious to discover a model of "sustainable" development, a way of protecting the forest while improving the lives of the people, that we might all (smugly?) subscribe to. What I actually found out is that no such model yet exists. The best that experts can provide for us is some kind of "goal", a framework of values yet to be filled with knowledge and experience.

Consider this. Everywhere in the Amazon the destruction of

the forest has followed the building of roads. The BR-364 has ripped through Rondonia, and there are plans to drive it on through Acre to Peru and the Pacific, opening the veins of the Amazon to the thirst of the Japanese for hardwoods and raw materials. Jose Lutzemburger, the celebrated environmentalist who is now Brazil's Minister for the Environment, has gone on record to say that the road will not go through. Surely, I thought, any self-respecting environmentalist must welcome this.

But no one I spoke to in Acre either believes or agrees with it. "There's nothing wrong with roads, you know", Macedo said to me. "It's who and what they are used for that matters." I think he's right. You can't condemn the people of Acre to live forever in the nineteenth century. If you do they will simply leave the forest and there will be no one to protect it. Banning roads in the Amazon doesn't quench the rich world's thirst for drugs and hardwoods, bring about agrarian reform or tackle the gross injustice of Brazilian society.

Again, the thickest smoke over the poorest part of Paragominas comes from the top of a hill. Fully prepared for yet another mental condemnation of the *patroes* (wealthy), what I actually found was a ragged-trousered collective of very poor people using the waste wood left by the sawmills to make charcoal. The operation was overseen by an old, tough, black woman wearing a baseball hat.

Should I applaud the collective or condemn the destruction? Close down the charcoal collective and you have a dozen or more destitute families ready to cut down the forest instead — and the wood still goes to waste. The point is that simple, sure-fire prohibitions in response to complex human situations can lead us to false conclusions that may actually make things worse.

Becoming "Engaged"

Many of the people I met were very anxious indeed to develop markets for "sustainable" products that don't damage the forest — and, above all, that can provide a decent income. Brazil nuts are one of the most important of these. So yes, go out and buy as many as you can from the co-operatives that produce them in Extractive Reserves. That means finding out about your nearest "alternative traders" who take a detailed interest in how the products they sell are produced, and buying from them, not from the supermarket.

The point is that in the process you have taken an active interest in what is going on, — become "engaged". And that is what everyone I met in the Amazon, without exception, wants you to do. You might go on from there to support an immediate ban on the trade in tropical hardwoods, which is urgently needed, or seek out beautiful Kampa handicrafts when and if they reach the market. And you might go on from that to wonder why your own government may be supporting the status quo in Brazil and demanding the blood of the Amazon in payment of its ill-gotten debts.

There would then, I think, be no shortage of obvious things for you to do, or of reasons for doing them. In supporting, say, groups that work for human rights and land reform in Latin America you would also be engaged in the struggle to save the Amazon forest, in bringing rich world concern for the environment together with Third World concern for human life.

For myself, I shall not forget sitting, tired, hot and hungry, in the middle of a babassu palm grove on land still owned by former President Sarney, one of the biggest landowners in Maranhao. Around me were people who had "invaded" it, talking about the violence they had endured and what they need now. Rather timidly I asked whether Brazil might one day be run, not by the hated *patroes* (wealthy), but by the poor. There was a pause. "Yes!" said the woman sitting next to me. Suddenly, quite unexpectedly, she smiled. Then everyone burst into laughter and applause.

13   SAVING THE FOREST: IDEAS IN CONFLICT

FOREIGN HELP TO SAVE THE FOREST

U.S. Agency for International Development

The U.S. Agency for International Development (USAID) is a federal agency committed to international social and economic development through direct funding and other assistance programs.

Points to Consider:

1. What is USAID's central objective?

2. How is USAID working to halt deforestation?

3. What is the vital forestry initiative?

4. How does USAID represent positive international efforts to save the forests?

USAID, "Tropical Forests: Saving a Vital Resource," USAID **Highlights**, Winter 1989.

USAID fights to conserve the world's tropical forests not only for their usefulness, but for their beauty and unmatched diversity.

Existing relatively undisturbed for millennia, tropical forests once covered more than four billion acres. Forests in some areas survived even the ice ages and are among the most ancient ecological systems on Earth. Although they now cover only 2.5 billion acres, seven percent of the Earth's surface, tropical forests are home to more than half the world's plants and animals — between 2.5 million and 20 million species. This biological diversity is unmatched anywhere else on the globe. But conservationists say that each day as many as 140 species of plants and animals may be disappearing as tropical forests are destroyed.

As forests are destroyed, the indigenous people who dwell in them are displaced, and their centuries-old knowledge of the forest and its uses may be lost. For example, more than 130 species of medicinal plants used by the Quichua and Shuar Indians of Ecuador have been collected, including treatments for diarrhea, fever, intestinal parasites and snakebite. The scientific community needs these resources and time to study them.

Recognizing the crucial importance of tropical forests and the plants and animals within them, the USAID is working to curtail destruction of forests and protect them by encouraging their use in a sustainable way.

"The Agency's central objectives of promoting economic expansion and improving human welfare in developing countries are critically dependent on the sustainable use of natural resources," USAID Administrator Alan Woods says. "USAID recognizes that the quality of life depends not only on economic and social development, but on environmental factors."

Protecting the Environment

The burning of tropical forests pumps nearly one billion tons of carbon dioxide into the air each year, significantly contributing to atmospheric gases thought to be causing a global warming trend (the greenhouse effect) by trapping the sun's heat in the atmosphere.

To many developing countries, however, cutting and clearing the forests may seem a necessity. Most are faced with the problems of feeding ever-growing populations, supplying lumber and fuelwood, and generating income to fund development and

Surveying the rain forests of the Philippines. Source: U.N. Food and Agriculture Organization

service massive foreign debts.

USAID's tropical forestry program has become an important and integral part of the Agency's overall development effort. USAID currently has $586 million (life of project) committed to approximately 150 forestry activities worldwide. The Agency also is exploring the possibility of debt-for-nature swaps (buying a portion of a developing country's forestry or other environmental projects).

To address the challenge of stopping deforestation, USAID has undertaken four interrelated approaches to conservation and management of tropical forests and their plant and animal life, including:

- protection of tropical forests and natural habitats;

- sustained production of plant and animal products and services from natural areas;

- rehabilitation and reforestation of cut-over and eroded areas; and,

- intensive management for improved plant and animal yields through agroforestry.

"Used in concert, these approaches can help developing countries improve rural income and wealth, conserve forest resources, maintain watershed and environmental services and protect biological diversity," says Jack Vanderryn, USAID director for energy and natural resources.

Setting Regional Priorities

Most agency-assisted countries in Africa have at least one forestry project or a forestry component of a larger project. In 1987, USAID launched a long-range natural resources management plan for sub-Saharan Africa. The plan makes natural resources management an important and integral part of all the Agency's African programs, including agriculture and rural development, women in development, policy dialogue, human resources and training, and data collection.

Of forestry projects in Asia and the Near East, three-fourths are aimed primarily at economically self-sustaining activities focusing on forest rehabilitation and intensive on-farm tree crop management through agroforestry.

In Latin America and the Caribbean, rapid deforestation and inappropriate land use have left hillsides eroded, making it difficult for poor farmers to produce sufficient crops for food or market. USAID's strategy in Latin America integrates protection and management of fragile lands primarily in tropical forest areas with sustained use for timber and other forest products as well as rehabilitation of eroded hillsides.

Vital forestry initiatives include:

- supporting preparation of country environmental profiles, and resource assessments and conservation strategies to help countries examine their environmental problems more carefully and to plan effective solutions;

- funding an International Board for Plant Genetic Resources study of wild relatives of domesticated crops;

- encouraging the Honduran Forestry Development Corporation, the Honduran government agency responsible for forest management, to protect forests and use them as a renewable resource;

- employing refugees to establish nurseries and reforest 12,000 acres in Somalia and eastern Sudan;

- contributing to establishing a gene bank in India that will preserve germplasm for use by scientists, as well as working with the World Bank and the government of India to generate sustainable forest production and increase rural incomes in four Indian states; and

- supporting Ecuador's *Fundacion Natura* in its campaign to gain support for conservation and sustainable development.

Promoting Forest Management

Deforestation on slopes and watersheds causes soil erosion, which leads to clogged waterways, reservoirs and irrigation canals; flooding; and disrupted hydroelectric projects. Half of the world's population lives in mountains or adjacent lowlands that are affected by tree cover on mountain watersheds. In 1988, deforestation of watersheds contributed to catastrophic floods in Bangladesh and Thailand.

About one-fourth of USAID's current portfolio of forestry projects focus primarily on land rehabilitation, reforestation, and natural forest and watershed management, exemplified by USAID's seven-year Forestry Land Use Planning Project in Niger begun in 1980. At the national level, the project has helped initiate development of a functional planning unit within Niger's Forestry Service that integrates land use planning and natural resource management with the country's other development activities.

14   SAVING THE FOREST: IDEAS IN CONFLICT

FOREIGN INTERESTS THREATEN THE FOREST

Sandra Steingraber and Judith Hurley

Sandra Steingraber and Judith Hurley originally wrote this article for Food First News, *a publication by the Institute for Food and Development Policy.*

Points to Consider:

1. What is at the root of hunger and deforestation problems in Brazil?

2. Who are the real victims of development?

3. How are the Brazilian government, police and courts to blame?

4. How are foreign investors and interests to blame?

This article is excerpted from a **Food First Action Alert**, Winter 1990, titled "Brazil's Debt and Deforestation—a Global Warning", authored by Sandra Steingraber and Judith Hurley.

With start-up funds provided by the World Bank, [one] project involves not only strip mining vast areas for iron-ore deposits, but also logging hundreds of thousands of additional square miles.

As we observe the 500th anniversary of Columbus' "discovery" of the Americas, the ongoing conquest of the New World is taking place in the Amazon region of Brazil. In the name of development, timber merchants, goldminers, impoverished migrants, cattle ranchers and foreign investors are conducting a massive foray into the largest remaining forested frontier in the hemisphere.

Less publicized are the root causes of environmental and human rights abuses as well as other related abuses whose impact extends far beyond the Amazon. Along with the disappearance of the actual forests, the damming and toxic contamination of the Amazon River tributaries threaten the world's largest fresh water system. Along with the indigenous people of the rain forest, 100 million other Brazilians are suffering acutely as the result of policies that impoverish their communities, turning many of them into agents of further environmental destruction. Who is responsible?

Neocolonialism

Brazil is the fifth largest country in the world, comparable in size to the continental United States. Considered on the basis of GNP, Brazil is by far the wealthiest country in the Third World.

Yet amid this great wealth, two out of three Brazilians go hungry. An estimated 1,000 children die from hunger-related causes every day. Thirty-six million children are ill-fed, ill-clothed and ill-sheltered, unable to attend school, and exploited as the most vulnerable members of the labor force. One child in 10 has at least one debilitating handicap, such as blindness, deafness or mental retardation, traceable to malnutrition in early childhood. If we look closely, we find that the causes of hunger and the causes of rain forest destruction are very closely related.

At the root of these problems is an identification of development with unlimited economic growth, modernization with foreign capital accumulation, progress with trans-nationalization and profit maximization, and satisfaction of real human needs with consumerism.

Much of the poverty in Brazil can be traced to the concentration of land ownership. According to a recent

101

Cartoon by Hubig, **Pacific News Service**

government report, the richest one percent of Brazilians receive as much national wealth as the poorest 50 percent. Less than four percent of landowners own 70 percent of the land. Each year one million of the highly productive small farmers lose their land to large landowners or to agricultural, industrial or financial corporations.

Most of the 25 million small farmers who have been pushed off their land during the past two decades have migrated to urban areas, creating the *favelas* (shantytowns) that surround every city. They arrive in desperate need of work, but lack urban skills.

Such large concentrations of dissatisfied people constitute a real threat to the status quo. Where was the Brazilian government to find a safe outlet for this smoldering discontent? Dividing up large agricultural estates as a means to agrarian reform could be too politically explosive. So instead, the government promoted colonization of the "underexploited" Amazon as the solution to landlessness.

With international aid, in the early 1970s, the government extended access to the rain forest by building the Transamazon

Highway, followed by a number of other roads. It carried on an intense publicity campaign among the poor, offering the inducement of free land, and painting the Amazon as the new frontier of opportunity. This campaign succeeded in attracting hundreds of thousands of the landless and unemployed.

But the impoverished settlers encountered anything but a utopia. When their crops fail—usually within two to five years due to soil exhaustion—settlers sell or simply abandon their land, moving deeper into the rain forest in search of new land.

The Real Victims

The groups most threatened by the wave of migration and subsequent deforestation of the Amazon are the rubber tappers and the Indians. To the rubber tappers—many of whom are descendants of the earlier migrants—agricultural settlement destroys the trees out of which they literally extract a living.

For indigenous people, these migrations constitute an invasion of their tribal lands. Indian territories are not clearly demarcated, so it was inevitable that conflicts would arise when settlers from the shantytowns arrived with titles to poorly defined land holdings.

More dangerous to peace and to the preservation of the Amazon than colonization, has been the government's policy of luring wealthy business into the rain forest with tax breaks and crop subsidies. In 1988, Indians from at least 10 groups were killed in Brazil for their efforts to protect their land against invasion by cattle ranchers and timber merchants.

In 1987, over a million people were involved in 715 documented land conflicts. Approximately 1,000 farmers, rural workers and their allies, including priests, nuns and lawyers, have been killed in land conflicts since 1980.

The police and the courts have been squarely on the side of the ranchers as the violence has escalated. In the past couple of years, paramilitary forces have targeted the most important leaders active on behalf of the rural poor, and death squads organized by landowners have already killed many on their well-publicized hit lists.

The murder of Chico Mendes in December 1988 drew world attention to the violence in the Amazon. Mendes was president of the Rubber Tappers Union of Xapuri in the jungle state of Acre.

Opponents of Reform

Opposed to reforms is the powerful Rural Democratic Union, a cattle ranchers' cartel that is also responsible for much of the violence against environmental advocates like Mendes. Formed in 1985 in response to the National Agrarian Reform Act, the union now has a considerable influence in the Brazilian Congress and a very strong presence in the mass media.

As a result of the union's activities, the new Brazilian constitution contains provisions on land tenure even less favorable to land reform than those in effect under the previous military dictatorship.

Many of the disenfranchised have meanwhile organized as the Movement of Landless Rural Workers, a rapidly growing organization that has carried out large-scale land occupations in several states. In some areas, the polarization between landowners and the landless approaches civil war. To understand the economic forces that create and maintain violent conditions in the Amazon, we need to look at Brazil's greatest economic pressure: the $117 billion foreign debt.

Foreign Debt

The International Monetary Fund (IMF) has intervened to renegotiate Brazil's debt. In order to maintain its credit standing, Brazil was forced to accept IMF policies that require the government to take out new loans to meet the payments on the original debt. Interest payments currently amount to approximately $12 billion per year—the equivalent of the minimum wage for 20 million Brazilian workers.

Other conditions imposed by the IMF called for government cuts in spending on health, education and housing; the elimination of subsidies for essential foods and public transport; the reduction of real wages; and the liquidation of job-creating state enterprises. As a result, the Brazilian worker, since 1980, has suffered a 40 percent loss in buying power. In 1987 alone, the cost of bread went up 653 percent.

These economic austerity measures, with their devastating effects on rural and urban people alike, in no way provided enough funds to offset the billions due on the debt. To acquire the necessary capital, the IMF imposed further conditions on Brazil, emphasizing greater export earnings through new business investments, including new development schemes.

One such scheme, mineral exploitation; is as devastating to the ecology and economy of Brazil's rain forest as colonization. These activities involve both megabusinesses and desperately poor individuals acting as the agents of destruction. They consume vast amounts of electricity and resources and leave a trail of toxic wastes in their wake.

The Greater Carajas Development Program is a $62 billion case in point. A mining project that occupies an area of eastern Amazonia the size of Britain and France combined, it has been labeled as "an investment in ecological and social disaster".

With start-up funds provided by the World Bank, the project involves not only strip mining vast areas for iron-ore deposits but also logging hundreds of thousands of additional square miles to provide charcoal fuel for the smelters. By the year 2010, the Brazilian government plans to build over 136 more hydroelectric dams along the Amazon waterways.

While grass-roots groups within Brazil pressure the government to bring about change, their popular leaders urge North Americans concerned about rain forest destruction to work to change the policies of the institutions that underwrite the status quo: the World Bank and the IMF.

As Brazilian economist Marcos Arruda has observed, what is required is not structural adjustment of Third World economies to First World interests, but structural transformation — the development of an economy based on human needs.

15

SAVING THE FOREST:
IDEAS IN CONFLICT

DEBT-FOR-NATURE SWAPS ARE BENEFICIAL

Nick Lenssen

Nick Lenssen wrote this article on debt-for-nature swaps for Worldwatch *magazine. Debt-swapping is an international scheme to help relieve Third World debt and preserve the environment in developing nations.*

Points to Consider:

1. What is a debt-for-nature swap?

2. How will a debt swap help the rain forests of Ecuador?

3. What are other benefits of debt-swapping?

4. How do recipient countries maintain control of their debt dollars?

Nick Lenssen, "Debt-for-Nature Swaps," **Worldwatch,** November/December 1988.

Swaps allow a debtor country to invest local currency in its own future rather than export scarce dollars to cover debt service.

It's become clear that debt problems and environmental deterioration have strained the ability of many developing nations to climb out of their economic hole. Throw in worsening poverty and high population growth rates, and the prospects for these countries ever to see daylight seem very dim.

Recently, an innovative method has evolved to provide funds for conservation activities while simultaneously chipping away at debt burdens. Although the sums involved are still marginal (a total of $72 million, compared with $1.2 trillion in outstanding debt), if successful this initiative could be expanded to provide a model for large-scale programs that show countries a way out of the debt and poverty trap.

The method is called the "debt-for-nature" or "debt-for-development" swap. Put simply, private organizations in industrial countries acquire Third World debt at discounted rates or by donation. This debt is then exchanged for commitments by the debtor country and local conservation groups to undertake projects that protect the environment or develop the local economy in a sustainable manner.

Although patterned on debt-for-equity exchanges—in which a corporation or other institution may use discounted debt to acquire foreign assets, such as a factory, at bargain rates—debt-for-nature swaps do not transfer ownership of assets to the industrial world. Instead, the proceeds from the deals are administered by local nonprofit groups in government-approved programs. All parties to the deal benefit, as does the environment.

The preservation of Yasuni National Park in Ecuador's Amazon forest offers a textbook case of how a swap works. Yasuni is threatened by oil exploration—and a proposed highway—but the government ministry in charge of the park lacks the funds to manage and protect it. That all may change as the result of a debt-for-nature swap.

In an agreement between the Ecuadorean government, the World Wildlife Fund (WWF) and Fundacion Natura—a nonprofit conservation group based in Quito—the WWF purchased $1 million of Ecuadorean debt on the open market for $354,500. In exchange, the Ecuadorean Central Bank issued bonds in Fundacion Natura's name for the full $1 million but in the local

107

A section of remaining rain forest in the Republic of Vietnam. Source: United Nations Food and Agricultural Organization.

currency. Fundacion Natura will use the interest income from the bonds to finance protection and management activities in six major parks—including the Galapagos Islands, as well as Yasuni—while the principal will be added to the Fundacion's endowment when the bonds mature in nine years.

A swap that goes beyond environmental protection was signed earlier this year, when the Dutch and Costa Rican governments reportedly agreed to retire up to $50 million in commercial bank debt in exchange for $15 million worth of local currency investments in reforestation, watershed management and soil conservation projects. Instead of locking away natural resources such as tropical forests—a criticism of past swaps—this accord will have Costa Rica invest in its agricultural and resource base.

108

Swaps allow a debtor country to invest local currency in its own future rather than export scarce dollars to cover debt service. They free local currency funds for projects that have a positive effect on the natural and human environments. And, these investments have the potential to strengthen not only programs in, say, reforestation or family planning, but also the often weak and unrecognized institutions that administer them—an important consideration in developing countries.

Swaps do more than force a perceptual connection between debt and today's global problems. By preserving tropical forests they are already providing a marginal amount of real investment in helping curb global warming and the loss of biodiversity. Unless the current international financial impasse is resolved, hope for future investments in these and other areas is bleak indeed. Swaps show a way to accomplish both.

16 SAVING THE FOREST: IDEAS IN CONFLICT

DEBT-FOR-NATURE SWAPS ARE NOT BENEFICIAL

Gus Yatron

Gus Yatron is chairman of the National Security and International Affairs Division of the U.S. General Accounting Office (GAO). The GAO is a federal agency that prepares special reports for the Congress on Budget policies and public issues.

Points to Consider:

1. What are PVO's?

2. Have debt swaps helped reduce Third World debt?

3. How are debt swaps a disadvantage to PVO's?

4. Why is sovereignty a factor?

Gus Yatron, General Accounting Office, Report to Congress, "Developing Country Debt: Debt Swaps for Development and Nature Provide Little Debt Relief," December 1991.

Debt swaps have not significantly reduced any country's external debt.

A debt swap is a form of debt conversion in which external hard currency debt is traded for local currency or debt denominated in local currency. Debt-for-nature swaps support specific environmental projects, such as the designation and management of protected areas, the development of conservation management plans, the training of park personnel, and environmental education activities.

From 1987 through 1990, 13 countries completed 26 debt swaps. These swaps retired debts totaling about $126 million (less than one-twentieth of one percent of the countries' external debt and less than one-fifth of one percent of their commercial debt). Of the $126 million, $86.4 million was exchanged by Costa Rica. Fifteen swaps-for-nature accounted for nearly 90 percent of the $126 million, while 11 swaps-for-development accounted for about 10 percent.

The advantages and disadvantages of debt swaps vary by participant. PVO (private voluntary organizations) participation in debt swaps enhances their ability to raise funds and offers PVOs the prospect of providing additional funding for the PVO's programs. However, many PVOs have not participated in swaps because of their complexity and the high resource commitment required.

The list of disadvantages is lengthy: the potential inflationary impact of debt swaps, the relatively high price that could be paid for the debt, the implicit subsidy provided to PVOs through a debt swap, the cost of servicing domestic bonds issued in exchange for the external debt, the perceived loss of sovereignty incurred if land ownership were transferred to an international PVO, and the potential for debt swaps to restrict remunerative development activities. Some economists believe that even if these disadvantages can be overcome, a debt swap makes sense only if a straight donation would not otherwise occur. USAID (U.S. Agency for International Development) has had limited involvement in debt swaps, viewing its role as more catalytic than direct.

The plan was to provide some debt relief to developing countries with debt-servicing problems in exchange for local currency resources that would be used to carry out environmental and development projects in these countries. This concept became known as debt-for-nature or debt-for-development swaps.

"The Law of the Jungle? Well, son—these days that's a little hard to explain."

In these swaps, a PVO uses its funds to purchase a debtor country's external commercial debt on the secondary market at a significant discount from the debt's face value. Then, the PVO and the debtor country usually exchange the external debt instrument for domestic currency or a domestic currency bond that the local PVO uses to fund agreed-upon projects. Generally, all terms of the swap are agreed to before its execution.

Advantages and Disadvantages

Although advocates often describe debt swaps as "win-win-win" situations in which all participants (the private voluntary organizations, commercial banks, and the developing countries) benefit, there are disadvantages for each party that participates.

A disadvantage to a PVO is the amount of time and staff resources it takes to finalize a swap agreement; some swaps have taken up to 18 months to complete. The negotiation process between a PVO, the debtor country, and commercial banks has almost always been complex and labor intensive. The costs of this complex process can limit PVO participation in debt

---

WORLD BANK INTERFERENCE

Guyana is one of the world's most indebted nations. Yet foreign development agencies have been pushing the country to intensify logging. In 1984, the World Bank's soft loan facility, the International Development Agency, pumped $8.8 million into the industry in an attempt to boost production. In 1989, under the UN's Food and Agriculture Organization-administered "Tropical Forestry Action Plan", the Guyanese Government, with the help of foresters from Canada, developed a new plan to rapidly expand the areas to be logged.

Foreign companies have begun to move into the forests in increasing numbers. A Venezuelan company, Palmaven, has gained a concession to exploit 300,000 hectares on the Demerara River, south of Linden.

Marcus Colchester, "Sacking Guyana," **Multinational Monitor**, September 1991

---

swaps, especially for smaller, inexperienced PVOs.

The most obvious advantages for a debtor country that participates in debt swaps are the environmental and development programs it receives and the reductions to its external debt. There are, however, potential disadvantages to debt swaps, such as (1) the relatively high price that could be paid for the debt, (2) the implicit subsidization of the swap, (3) its potential inflationary impact, (4) the financial cost of servicing the domestic debt issued in exchange for the external debt, (5) the perceived sovereignty issue, and (6) the potential for restricting resource development, which could dissuade developing countries from participating in a debt swap.

One of the more controversial aspects of debt swap arrangements, particularly when this mechanism was first used, was the perception among some people in developing countries that their country might lose some of its sovereignty as a result of the transaction. For example, some people were concerned that if land ownership were transferred to an international PVO, the country would no longer have control over this resource. While such transfers have not occurred to date, the perception persists and is a serious concern in some debtor countries.

SAVING THE FOREST: IDEAS IN CONFLICT

A "NEW FORESTRY" IS THE ANSWER

John C. Ryan

John C. Ryan is a researcher at the Worldwatch Institute *where he special- izes in forest issues. This reading deals with adopting a "New Forestry" approach in managing the rain forest.*

Points to Consider:

1. What is meant by a "new forestry", and why is it needed?

2. Describe the concept of sustained yield.

3. Why are native forests better than "monocultures" and tree plantations?

4. How is Peru's Palcazu Project minimizing the effects of deforestation?

John C. Ryan, "Timber's Last Stand", **Worldwatch**, July/August 1990.

Given that logging of primary forests is not going to stop tomorrow, New Forestry promises to minimize the damage to areas that will be lumbered.

The earth shakes as the forest giant crashes down, its crown landing hundreds of feet from the massive stump left behind. The product of countless seasons of clean air, stable climate and freedom from human disturbance, the centuries-old tree represents a windfall profit to its sawyers, one never to be realized again.

Whether it was a western hemlock or a Philippine mahogany, the tree's fall also symbolizes one of the most widespread, and visibly shocking forms of environmental degradation: deforestation. Few images breed concern for the planet like the denuded moonscape of a clear-cut plot of scorched earth that was once rain forest.

It is clear that the time for a new forestry—one that seeks to use and maintain the complexity of forests, rather than eliminate it—is now. Although it's hard to imagine a timber industry that treads lightly on the land, it should be possible, since wood, in theory at least, is a renewable resource. As human numbers and needs continue to rise, and forests continue to dwindle, it is urgent that we learn to tap forest's varied riches without impoverishing their source. As the timber industry confronts the exhaustion of both woodlands and the public's tolerance of its practices, it faces the inevitable choice of living up to this challenge, or following its resource base into rapid decline.

Fortunately, in places as disparate as Peru's Palcazu Valley, small bands of researchers and activists are piecing together the beginnings of a sustainable forestry.

The Economics of Destruction

As the world's greatest storehouses of life, forests are valuable for much more than their timber. When these riches are sacrificed to wood production, logging often becomes difficult to justify on economic grounds.

The costs of logging have usually fallen on those who depend on intact forest—forest dwellers, downstream communities, tourism-based economies, among others. But as the area of untouched forest shrinks, it is becoming clear that the timber industry is also putting itself out of business. Tropical hardwood exports, worth $8 billion in 1980, have fallen to $6 billion, and are projected to shrink to $2 billion by the end of this decade.

Trees for income—the timber industry in Peru. Source: U.N. Food and Agriculture Organization

When diverse populations of trees are replaced with genetically uniform stands, there is a double loss to future timber harvests. Plantations can relieve pressure on natural forests by producing wood quickly, but because the natural system of checks and balances has been stripped to maximize tree growth, monocultures (one crop rural economics) are prone to unravel. (A serious crop disease can destroy the whole one crop economy)

When native forests are lost, industry also loses its reservoirs of genetic variety and its scientific laboratories for uncovering the many hidden relationships that make timber growth possible. For instance, according to work done by Oregon State University entomologist Tim Schowalter and others, intact stands of natural forest are valuable as physical barriers and as sources of insect predators to stop the spread of pest outbreaks on adjacent plantations. As long as wilderness is left to tap and study, then foresters have the opportunity to learn from their mistakes. But, as industry converts native stands to plantations, its options keep narrowing.

Toward a New Forestry

How can timber be harvested without destroying forests? The answer to this question is being discovered in some very unlikely places: in a chunk of rotting wood on a forest floor, amid a buzz of insects hundreds of feet up a Douglas fir, in the fecal pellets of a flying squirrel. In these and untold other places lie the essentials of forest productivity that foresters ignore to the detriment of forests and timber production.

A small group of researchers and forest managers based at the H.J. Andrews Experimental Forest in western Oregon have been studying the lessons of natural forests and have started applying them in an attempt to reconcile the seemingly unsolvable conflict between logging and forests. The "New Forestry," as their ideas are being called, represents a fundamental change, a revolution, even, for the forestry profession, which has traditionally focused narrowly on timber production.

Foresters are starting to recognize dead trees and logs on the forest floor as essential parts of a healthy forest, not a form of waste to be burned off or shredded. Besides providing important wildlife habitat, woody debris maintains soil fertility by returning organic matter and nutrients to the soil and by helping to control erosion.

New Forestry is no substitute for protecting natural areas: no forester can recreate 1,000-year-old ecosystems or bring back species driven to extinction. Environmentalists are rightly suspicious of anyone trying to sell new types of logging as the solution to deforestation. Especially where the amount of wilderness left is small, preservation is still the top priority.

Nonetheless, given that logging of primary forest is not going to stop tomorrow, New Forestry promises to minimize the damage to areas that will be lumbered. New Forestry has been researched and applied almost exclusively in the ancient forests of the Pacific Northwest, but, as David Perry, a forest ecologist at Oregon State University, notes, although particular techniques will vary greatly, "there are certain ecological principles translatable virtually anywhere in the world."

Reducing the risk of future pest outbreaks and ensuring that soil is not robbed of its nutrients makes sense whether wood is harvested from a pristine rain forest, a logged-over woodland, or an intensively cropped tree farm. As evidence builds that the intensive forestry practices used today often fail to sustain timber productivity over time, timber managers may see the wisdom of

---

WHAT WILL INHERIT THE EARTH?

The destruction of tropical rain forests is killing off at least 4,000 wild species a year—and top scientists are calling for a halt to new development. "One-quarter or more of the species of organisms on Earth could be eliminated within 50 years," if current rates of clearing continue, they say. A pretty planet it won't be.

Only the hardiest and most adaptable of plants and creatures are likely to be around in 200 or 300 years, say scientists. They include: flies, weeds, starlings, grackles, rats, raccoons and, you guessed it, cockroaches. The cockroache, the creature that has endured for millions of years despite the best efforts of homeowners and apartment dwellers, will probably last millions more, whatever havoc we wreak.

USA Today, "The Dire Results of Rain Forest Destruction", August 16, 1991

---

restoring natural resilience and diversity to their lands.

Tropical Troubles

In primary rain forests, still the predominant resource for the tropical timber trade, the social, political and biological complexities of forest use raise doubts whether sustainable timbering is even possible. Removing too much wood from these forests, in which nutrients are found mostly in the plant life itself, not in the soil, leaves behind a nutritionally impoverished system that may take hundreds of years to rebound. Even selective logging is atypically very destructive because of the tremendous diversity of tree species: loggers inevitably trample wide areas as they "cream" the forest—taking only a handful of desired species.

Third World governments, saddled with debt and swelling populations, typically see their rain forests as quick sources of foreign exchange or as safety valves for an expanding underclass. Unstable conditions outside the forest make long-term policy inside—such as enforcing minimum lengths of logging rotations or preventing illegal entry in logged-over areas—difficult to enforce. The sheer number of people looking to tropical moist forests as sources of sustenance and profit may already overwhelm their carrying capacity.

Despite the numerous obstacles, a handful of projects show how sustainable tropical logging might work. The Yanesha

Forestry Cooperative, the first Indian forestry cooperative in Amazonia, has been operating since 1985 in Peru's Palcazu Valley. Local people own and process the forest products; timber cutting is designed with protection of diversity in mind. By clear-cutting in narrow strips, leaving most of the forest intact, the Palcazu project seeks to mimic small-scale natural disturbances. Creating gaps in the forest canopy allows the shade-intolerant seedlings of hundreds of different species from the uncut areas to colonize the strips. Bark and branches are left in place to maintain soil fertility, rather than burned off.

New approaches to logging that incorporate rather than ignore natural linkages and local people can, at the very least, prolong the useful life of logging areas, and buy time for other solutions to deforestation to be worked out.

Protecting Forests, Protecting Jobs

While the crush of human demands ensures that most forest will be used in one way or another, turning natural areas of global significance into pulp — as is happening in Alaska's Tongass rain forest and Southeast Asia's last large area of coastal mangroves in Bintuni Bay, Papua New Guinea — is a travesty considering that pulpwood can be obtained with much less impact from second-growth forests and plantations. Ending subsidies for conversion of primary forests and putting in place incentives for better management on less valuable lands could help increase sustainable timber supplies.

But new supplies will take time to develop; forests will continue to be pushed beyond their limits until the world begins to curb its spiraling appetite for wood products. Jobs and profits based on ecological destruction simply cannot last. If societies can come to grips with this fact, perhaps we can make the transition to sustainability while there are still ecosystems left worth protecting.

SAVING THE FOREST:
IDEAS IN CONFLICT

LAND REFORM AND SOCIAL JUSTICE IS THE ANSWER

David Ransom

David Ransom is co-editor of the Canadian publication The New Internationalist. *This article deals with issues of social justice in Brazil's Amazon rain forest.*

Points to Consider:

1. How are women treated in Brazilian society?

2. Why did the author find conditions so sickening?

3. What single measure will help halt deforestation the most? Why?

4. How are people organizing and joining together to ease social injustice?

David Ransom, "Urban Exiles", **The New Internationalist,** May 1991.

The single measure more likely than anything else to halt the continued deforestation of the Amazon is the equitable distribution and more efficient use of the land.

"People are stupid!" says Lucia, sitting in a cafe in the great city of Belem — "Bethlehem" — at the mouth of the Amazon.

"They don't want to see the benefits that unity can bring us. They are fooled into seeing only the things that divide us. But of course all of us poor people, black, brown, yellow, white, whatever race we belong to, whatever sex we are, all of us have many problems in common." The woman next to her nods her agreement.

The cafe is built over an open sewer in the industrial part of Belem. Across the road more women are gathered around a factory gate. It is one of the factories owned by the Mutran family, who control a large part of the city's Brazil nut processing industry. I have been brought here by Igina, a smart young woman who gave up commercial work to join the Movement of Rural and Urban Women.

Plenty of Machismo

All the women I talk to tell much the same story. They work long hours for minimal pay in unhealthy conditions.They are paid piece-work rates and receive nothing for damaged Brazil nuts although they are still sold in their crushed form. Pregnant women are sacked, and since many have half a dozen children or more, that adds up to an awful lot of sacking.

Most of the women live across the open sewer in a rambling shantytown that was "invaded" by workers in the Brazil nut factories 15 year ago. In some ways it is a familiar Latin American urban scene. But what makes this different is that all the workers in the factories are women and the community is largely run by them. They have joined the Food Workers Union, and some have become active in the Movement of Urban and Rural Women.

"As you can imagine, there's plenty of *machismo* in the union and plenty of members expect the women just to do what they are told," say Igina. "In fact there's just one woman full-time official. But unions in Brazil are only now beginning to recover from the inheritance of the military dictatorship when everyone had to belong to official unions. Their job was to keep their members in line. Now, for the first time, the leaders are beginning

A seedling nursery in Nicaragua. Source: U.N. Food and Agriculture Organization

to think about the interests of their members. They have even supported strikes by the women in these factories. Radical women are beginning to make an impact, and some of the men don't like that."

Violence and Poverty

These, too, are the people of the Amazon. They live in the cities and towns where once the forest stood and now the forces invading the Amazon are generated. This is where the people who quit the forest end up—and from where those who invade it frequently set out.

There is a point south of Belem that is known as the "Parrot's Beak"—because of the shape formed by the boundaries of the states of Para, Tocantins and Marnahao where they meet. Here the Belem-to-Brasilia highway crosses the new 900-kilometer railway running from Sao Luis on the coast to the giant Carajas iron ore mine. It should be blessed by the benefits of modern, industrial development. Instead it is the scene of the most intense and violent land conflicts in Brazil.

Take a look at Acailandia, perched on hilltops near the actual junction of the highway and railway. On these hilltops, balancing

MURDER IN THE AMAZON

In the Conceicao diocese and one adjoining township, more than 170 people have died in land conflicts from 1980 to now; most of the victims are small-scale agricultural laborers killed by military police or hired gunners, according to statistics. Conceicao holds Brazil's record for deaths in land conflicts, added Father Ricardo Rezende, a Catholic priest who started in the area as a lay activist in 1977.

One government official asserted that the annual homicide total reaches 300 for Conceicao and adjacent areas. Only one murder has come before a jury since 1981, the official added. "Killing occurs, and there isn't even a police report," he said.

Ken Serbin, "Earth Awaits Amazon's Ecological Showdown", **National Catholic Reporter**, December 14, 1990

precariously over eroded gorges, thousands of people live in utter squalor. Most of them are rural workers driven from the land owned by the *fazendeiros* (the ranchers). Without facilities of any kind, they live in minute wooden shacks. The smallest of all are on a patch of land "donated" by the friend of a local politician who made as many promises as possible in exchange for votes in recent elections. The air is thick with the acrid smoke of burning wood from hundreds of adobe ovens making charcoal.

Such conditions are common throughout the world. I've seen plenty just as bad. But there is something peculiarly sickening about this place. I try to figure out what it is.

Part of it is simply the sight of the great rain forest being destroyed. There are said to be some 600 sawmills in the 300-kilometer stretch between Imperatriz and Paragominas. They export the best hardwoods to the rich world. Second-class wood is used for the domestic building and furniture trades. The mills are incredibly wasteful, losing as much as 70 percent of the usable wood.

But the worst part of it is, I think, the way deforestation combines with massive foreign investment to produce utter human misery. One might argue forever about why this should be so. But you have to begin by seeing that it is so. And the biggest single complaint I hear from the people actually experiencing it is that no one wants to know. Talking to the active members of trade union and community groups, who risk their lives for their work, I discover what I think lies at the heart of my growing sense of outrage.

Land Reform

The single measure more likely than anything else to halt the continued deforestation of the Amazon is the equitable distribution and more efficient use of the land already cleared. That means land reform. It also means orchestrated violence and intimidation to prevent it, or any other challenge to the power of the wealthy.

Our companion, Gil, is a trade unionist who worked on the Belem-Brasilia highway. He stops as we drive along it and points to a place where he unearthed the bodies of fellow construction workers buried in a shallow grave. A little further on he shows us where laborers were simply burned to death in the forest instead of being paid. (The first news I receive on returning to England is that Ribeiro de Souza, president of the Rural Workers Union in Rio Maria, just to the south, had been assassinated. He was the twelfth rural workers' leader killed in the Parrot's Beak in just one month.)

The violence is directed precisely against those people who are leading the search for a better future, not just for themselves but for the forest as well. We talk to union leaders from the sawmills in Paragominas, and they are perfectly well aware that their work is destroying their children's future. Most of them are farmers who look for inspiration to the Movement of Landless People and the Union of Rural Workers.

The Unions

A sense of exile and displacement pervades the towns and cities of the Amazon. Indeed, it pervades urban Brazil. The greatest migration of all is, in fact, from rural to urban life—to the gigantic industrial conglomeration around Sao Paulo in Southern Brazil, where now upward of 40 million people live. And it is here that the Union of Indigenous Nations (UNI) works to strengthen the political influence of Brazil's indigenous people—and to develop alternatives to deforestation.

Ailton Krenak is the Union's national leader. His people, the Krenak, were evicted from their traditional lands before he was born. His parents moved to Sao Paulo. Perhaps it was a sense of displacement that gave him his early insights. "As a boy I already had a clear vision, but I had no real understanding," he says. "I have always had the feeling that the Indian nations are really one people."

So he set out on a voyage around Brazil to meet them, hitching lifts where he could, traveling with Indian chiefs to meetings. "It was as if the original streams were meeting to become one river, gathering strength as it flowed. We began to create expectations where before there had been nothing but despair."

Links have been forged with rubber tappers through the Forest People's Alliance, and Ailton is interested in working with rural and industrial unions. "But we're not interested in just talking. We need to take practical steps, small measures that will provide a positive example."

I travel with him to the Centre that the Union of Indigenous Nations has set up near Goiania on what was once forested land just south of Brasilia. Representatives of indigenous people from right across Amazonia come here to study alternatives to deforestation. Ailton wants them to return to their villages with the knowledge they have gained.

In the evening I go for a walk with Mario Krenak, the caretaker of the Centre. He takes me into a plantation of maize. At its center is a tiny hut. Inside it are Joveno, Maria and their two small children. They offer us food and coffee. They are landless migrants from southern Brazil who have been taken on by the Centre, though they are plainly not Indian.

We sit in silence, the "Indian", the "migrants" and the "gringo" together, watching the sun set over the valley as the children play with a plastic scooter. What I am finally beginning to realize is that no amount of intimidation can remove entirely from these people their sense of dignity or their courage to resist. The violence they have all endured can also take them beyond their personal suffering, across racial, cultural and geographical divides. That, I feel certain, is how their future, and that of the forest itself, will eventually be secured.

EXAMINING CULTURES AND COUNTERPOINTS

The ecocide of the world's vast tropical rain forests is being accompanied by a genocide of its human inhabitants, the indigenous populations who have made the forest their home for centuries. Before the encroachment of the "civilized" world, these native people developed cultures uniquely adapted to survival in the forest. From the Onge of India and the Mbuti of Africa to the Indians of the Amazon, these people have lived in harmony with their environment and are now faced with the loss of their lands, their culture and their very lives.

A. Understanding Indigenous Cultures

The readings in this book introduce us to the cultures as well as the struggles of the people in the rain forest. In Part A of this activity, describe an example from the readings that illustrates each point below.

1. The use of natural medicines
2. Controlled burning of the forest
3. Hunting, gathering and fishing for food
4. Religious and spiritual beliefs
5. Knowledge passed on
6. Examples of social behavior
7. Agricultural techniques

B. Examining Ecocide in the Rain Forest

In this part of the activity, we will compare the demise of today's global forests and native people with that of North American Indians and their land in the 19th Century. Read the statements below. Then read the instructions that follow the statements.

1. Roads and railroads cut through the American wilderness and brought thousands of settlers into traditional lands of the Iroquois Confederacy in Northeastern U.S.

2. The discovery of gold in the Black Hills of Dakota in the 1870s brought about the displacement of the Sioux and Cheyenne from their sacred lands.

3. Smallpox, introduced by the Europeans, decimated entire villages of Indians across America.

4. After the U.S. Civil War, the killing of over 10 million bison (buffalo), denied food, clothing and shelter to the Plains Indians.

5. Disease, starvation and murder took their toll on the displaced Indians of the Southeast U.S. during the "Trail of Tears".

6. Massive timber operations and extensive agriculture all but destroyed the culture of the Woodland Indians in America.

Guidelines for Part B and the statements above

1. For each statement above, locate a similar example from the readings in this book that deal with the oppression and the demise of today's indigenous people and the rain forests where they live.

2. You may also use examples from your own knowledge and other sources than this book.

C. Examining Counterpoints

THE POINT

For thousands of years the Indians of the Amazon have considered the rain forest their home. They have lived in harmony with the natural world and have contributed to its vitality and biodiversity. Governments must recognize the rights of these indigenous people to their land and should prevent further encroachment by miners, loggers and homesteaders. The onslaught of slash-and-burn agriculture, destructive mining practices and unrestrained logging operations that threaten both human and natural resources of the rain forest must come to a halt.

THE COUNTERPOINT

The population of tropical countries will increase to 1.5 billion by the year 2000. As the amount of arable land becomes more scarce, the rain forests must continue to be a valuable source of agricultural production and industrial development. In addition, many of these nations are burdened by tremendous foreign debt and must look to their natural resources as a means to build their economies. Logging, mining and other industries provide much needed revenue. Ecologically safer methods for development must be found along with a fair system of land ownership.

Guidelines

Social issues are usually complex, but often problems become oversimplified in political debates and discussions. Usually a polarized version of social conflict does not adequately represent the diversity of views that surround social conflicts.

1. Examine the counterpoints above.
2. How do both positions present reasonable claims on the rain forest?
3. Why will a solution to land use in the forest be difficult?
4. Write down other possible interpretations of this issue than the two arguments stated in the counterpoints above.

BIBLIOGRAPHY

GENERAL

Bathgate, D. Blame government mismanagement. *USA Today*, v. 119, June 1991: p. 6-7.

Carpenter, B. Faces in the forest. *U.S. News and World Report*, v. 108, June 4, 1990: p. 63-66.

Cassady, J. Rain forests may yield disease cures. *USA Today*, v. 119, June 1991: p. 5.

Castner, J. L. Pay your respects to the rain forest. *Sierra*, v. 75, Mar./Apr. 1990: p. 82.

Cunningham, A. B. Indigenous knowledge and biodiversity. *Cultural Survival Quarterly*, Summer 1991: p. 4-8.

Dawidoff, N. Doomed to die? *Sports Illustrated*, v. 72, Apr. 9, 1990: p. 84.

Diamond, J. M. Bach, God and the jungle. *National History*, Dec. 1990: p. 22.

Dufour, D. L. Use of tropical rain forests by native Amazonians. *BioScience*, v. 40, Oct. 1990: p. 652-659.

Elisabetsky, E. Folklore, tradition or know-how? *Cultural Survival Quarterly*, Summer 1991: p. 9-13.

Ellis, W. S. Rondonia: Brazil's imperiled rain forest. *National Geographic*, v. 174, Dec. 1988: p. 772-817.

Gannon, R. High-wire act in the rain forest. *Popular Science,* v. 238, June 1991: p. 86-91.

Griffiths, M. Hard days journey into Eden. *International Wildlife*, v. 21, May/June 1991: p. 44-51.

Halesworth, P. Plundering Indonesia's Rain forests *Multinational Monitor,* Oct. 1990: p. 8-15.

Hecht, S. and A. Cockburn. The fate of the forest—Developers, destroyers and defenders of the Amazon. *Verso*, 1989: 266 p.

Herer, J. The forgotten history of hemp. *Earth Island Journal*, Fall 1990: p. 35-38.

It's our rain forest. *Mother Jones*, v. 15, Apr/May 1990: p. 47.

Joyce, S. Snorting Peru's rain forest. *International Wildlife*, v. 20, May/June 1990: p. 20-23.

Larmer, B. The rain forest at risk. *Newsweek*, v. 118, Apr. 12, 1991: p. 42.

Lemonick, M. D. Hot tempers in Hawaii, *Time*, v. 136, Aug. 13, 1990: p. 68.

Mardon, M. Maneuvers in the teak wars. *Sierra*, v. 76, May/June 1991: p. 30.

Mardon, M. Steamed up over rain forests. *Sierra*, v. 75, May/June 1990: p. 80-82.

Matson, P. A. and P. M. Vitousek. Ecosystem approach to a global budget. *BioScience*, v. 40, Oct. 1990: p. 662-72.

Maxwell, K. The tragedy of the Amazon. *The New York Review of Books*, v. 38, Mar. 7, 1991: p. 39-48.

McIntyre, L. Urueu Wau Wau Indians: Last days of Eden. *National Geographic*, v. 174, Dec. 1988: p. 772-817.

Merwin, W. S. Bulldozing Hawaiian rain forests to fuel development. *Earth Island Journal*, Winter 1990: p. 22-23.

Morell, V. Rain forests: A genetic storehouse goes up in flames. *International Wildlife*, v. 20, Mar./Apr. 1990: p. 6-7.

Myers, N. Tropical deforestation: The latest situation. *BioScience*, v. 41, May 1991: p. 282.

Natural Resources Defense Council (NRDC). "The rain forest Book," *NRDC Publications*, 1990.

Opheim, T. Saving native Hawaii. *Utne Reader*, Mar/Apr 1990, p. 28.

Perera, V. Guatemala guards its rain forests. *The Nation*, v. 253, July 8, 1991: p. 54-56.

Rain forest priorities. *World Monitor*, v. 4, Aug. 1991: p. 10.

Repetto, R. C. Deforestation in the tropics. *Scientific American*, v. 262, April 1990: p. 36-42.

Revkin, A. The Burning Season. *Houghton Mifflin Publishing*, 1990: 317 p.

Rogge, B. A. Saving a forest: What can we do? *The Freeman*, May 1990: p. 169-70.

Ryan, J. C. Plywood vs. people in Sarawak. *Worldwatch*, Jan./Feb. 1991: p. 8-9.

Schwartz, D. M. Drawing the line in a vanishing jungle. *International Wildlife*, v. 21, July/Aug. 1991: p. 4-11.

Shoumatoff, A. The world is burning. *Little, Brown Publishers*, 1990: 317 p.

Smith, N. J. H. Conserving the tropical cornucopia. *Environment. 33, July/Aug. 1991: p. 6-9.*

Sussman, R. and G. Green. Saving Madagascar. *The Futurist,* v. 24, Nov./Dec. 1990: p. 43-44.

Tropical-forest "tragedy". [World Resources Institute Report]. *Popular Science,* v. 237, Sept. 1990: p 26.

Vanishing treasure. *National Geographic World,* v. 184, Dec. 1990. p. 22-27.

Wallace, J. Rain forest Rx. *Sierra,* v. 76, July/Aug. 1991: p. 36-41.

Weisenthal, D. B. The forest as a human right. *The Progressive,* v. 54, April 1990: p. 14-15.

Yancey, P. A voice crying in the rain forest. *Christianity Today,* v. 35, July 22, 1991: p. 26-28.

BRAZIL

Babbitt, B. E. Amazon Grace. *The New Republic,* v. 202, June 25, 1990: p. 18-19.

Brazil's rain forest murder trial. *Newsweek,* v. 116, Dec. 17, 1990: p. 38.

Browder, J. D. Extractive reserves will not save the tropics. *BioScience,* v. 40, Oct. 1990: p. 626.

Gould, J. J. Capitalistas in the mist. *Bussworm,* v. 3, July/Aug. 1991: p. 20.

Hayden, T. Rain forest journal. *Rolling Stone,* June 27, 1991: p. 51-53.

Opheim, T. Can Brazil nut ice cream save the Amazon? *Utne Reader,* Jan./Feb. 1990: p. 20-21.

Rabben, L. Brazil's military stakes its claim. *The Nation,* March 12, 1990: p. 341-2.

Serbin, K. Earth awaits Amazon's ecological showdown. *National Catholic Reporter,* Dec. 14, 1990: p. 15-22.

Shukla, J. and P. J. Sellers. Amazon forest unlikely to rise from ashes. *Science News,* v. 137, March 17, 1990: p. 164.

Smith, G. Amazon parable, *U.S. News and World Report,* v. 109, Dec. 24, 1990: p. 18.

Willrich, M. Murder in Acre. *The Amicus Journal,* Spring 1989: p. 10-13.

Worcman, N. B. Brazil's thriving environmental movement. *Technology Review,* v. 93, Oct. 1990: p. 42-51.

APPENDIX

ORGANIZATIONS FOR MORE INFORMATION

Environmental Defense Fund
257 Park Ave. S.
New York, NY 10010

Cultural Survival Inc.
11 Divinity Ave.
Cambridge, MA 02138

Rain Forest Action Network
301 Broadway
San Francisco, CA 94133

Conservation International
1015 18th St. N.W.
Suite 1000
Washington, D.C. 20036

Natural Resources Defense Council
P.O. Box 1400
Church Hill, MD 21623-9919

The Wilderness Society
900 17th St. N.W.
Washington, D.C. 20006

Sierra Club
730 Polk St.
San Francisco, CA 94109

World Wildlife Fund
1250 24th St. N.W.
Washington, D.C. 20037

New Forests Project
731 8th St. S.E.
Washington, D.C. 20003

American Forestry Association
P.O. Box 2000
Washington, D.C. 20013

Natural Resource Defense Council
40 West 20th St.
New York, NY 10011

The Nature Conservancy
1815 N. Lynn St.
Arlington, VA 22209

Earth Island Institute
300 Broadway
Suite 28
San Francisco, CA 94133